Contents

Typeset by Jonathan Downes,
Cover and Layout by SPiderKaT for CFZ Communications
Using Microsoft Word 2000, Microsoft Publisher 2000, Adobe Photoshop CS.
First published in Great Britain by CFZ Press

CFZ Press, Myrtle Cottage, Woolsery, Bideford, North Devon, EX39 5QR

© CFZ MMXIII

ISBN: 978-1-909488-13-7

Faculty of the Centre for Fortean Zoology

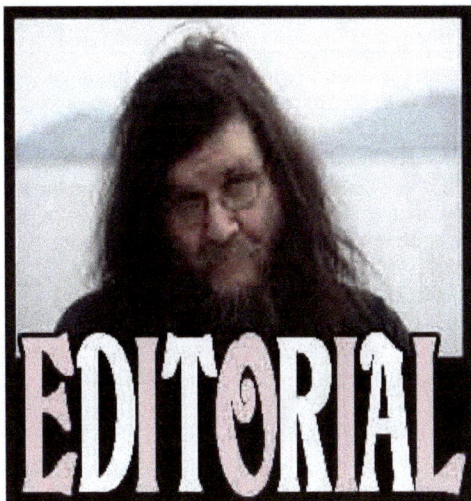

Dear Friends,

Welcome to another issue of the world's longest standing Fortean Zoological publication. Wow! Two in one year! That is something, by the standards of recent years. However, we do hope that as of 2014 we shall be back to the original schedule of three or four magazines a year. It is not that there is a shortage of material to put into these volumes; it is time that is at a premium, and as I get older and my life gets progressively more complicated, there seems to be less and less time for me to actually *achieve* anything.

The biggest cryptozoological bones of contention in recent months are those presented by Dr Melba Ketchum, and Professor Bryan Sykes. And before we go any further I would like to digress a bit by talking about something that Professor Sykes said in the opening parts of his current series on Channel 4. I am quoting from memory, but he - quite rightly - said that when cryptozoologists are claiming that "science rejects cryptozoology" they are talking nonsense.

Science is about answering questions, and so cannot be accused of rejecting cryptozoology out of hand. I wince whenever I read such an assertation on a cryptozoological website, which is something that happens all too often.

What they actually *mean* to say is that *MAINSTREAM SCIENTISTS* oh so often do not accept cryptozoological claims as valid.

I have had this happen to me. My assertions about the existence of pine martens and green lizards in southern England, both of which have been borne out as being perfectly true by the passage of time were - on several occasions - rejected quite rudely by scientists who really should have known better.

I understand *why* these scientists behaved as they did, but I do not forgive them for it. Somewhat easier to forgive are the scientists who are more receptive to cryptozoological ideas but who have been scathing about the state of cryptozoology as it stands in the early years of the 21st Century. That is because cryptozoology, if one looks around with a dispassionate eye, takes a deep breath, and accepts the unpleasant truth, is in a very dubious state.

Sharon Hill of *Doubtful News* recently wrote of a Bigfoot convention that:

> There is no critical thinking allowed. These events are for believers to share their experiences. Sound like a religious gathering? Yeah, it does for good reason. Many people who believe Bigfoot is in their back yard have zero solid evidence that it does but choose to believe out of faith and they wish it to be true.

I couldn't agree more. This is the main reason why I am no longer involved in UFO Research. It was all *I WANT TO BELIEVE* and precious little *I WANT TO FIND OUT THE TRUTH*. If I can quote another *X Files* aphorism, *THE TRUTH IS OUT THERE*. Yes, indeed, but it is quite often not what people want it to be, or even think that it might be.

I started the CFZ 21 years ago, because I wanted (amongst other things) to establish a culture of

The Great Days of Zoology are not done!

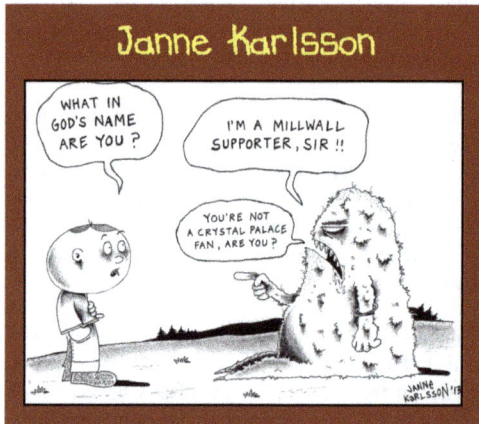

excellence, tolerance and mutual respect within the continually warring cryptozoological communities. In recent years, we have - I believe - come closer to achieving that aim.

I think that one of the biggest problems that there is about having cryptozoology respected by scientists, is that many (if not most) of its proponents do not behave in a scientific manner. To quote Sharon again:

> The name is "Squatchin' club"? REALLY? Ugh. Thanks *Finding Bigfoot* for a truly ridiculous name and a juvenilization of this activity. Yep, Bigfootery needs more silly words to bolster credibility.

She was writing about a Bigfoot Group in - I believe - Ohio, but the same could be said about many self-styled cryptozoological groups across the world. The current rash of crypto-themed television programmes in the West is - I would postulate - responsible for causing more damage to the concept of cryptozoology as a serious discipline than anything else in recent years.

But what has this got to do with Dr Ketchum and Professor Sykes? Many of us had high hopes from them, but - unfortunately - neither of them have actually done what I was hoping that they would set out to do. Unfortunately, although Dr Ketchum presented her paper as peer-reviewed, it appeared in a journal that she was apparently the owner of, and contained some remarkable

anomalies (which I went through in depth lat issue, and will not repeat here. But she published.

Professor Sykes hasn't. Which is a terrible shame. I have a lot of respect for Professor Sykes, and he and his team should be congratulated for coming up with evidence that there is a very strange species/subspecies/hybrid species/race of bear wandering about in the Himalayas. Their results are impressive, but they should have published them in a peer-reviewed journal before going public with them in a prime time TV show.

That at least partially negates the whole affair.

I would also argue that Professor Sykes and his team have *not* totally exploded the "myth" of the yeti as an unknown hominin. Remember also that of the witnesses cited, Reinhold Messner (who looked remarkably like a Jethro Tull roadie) always believed that what he had seen was a bear. The French explorer believed that the carcass from which he took hair samples was a bear, and the badly stuffed Nazi "yeti" was definitely a bear, so the people who thought that they had encountered a bear actually did so! The only local 'witness' interviewed hadn't actually seen the creature for himself, and was just working off a theory of what might have killed his livestock. So, as I said, congratulations to Bryan and the team - a new, or partly new bear species or subspecies is no mean feat.

However, to extrapolate from this evidence that there is no such thing as a mystery ape in central Asia, especially when any cryptozoologist worth his salt will tell you that these creatures are hardly ever seen above the snowline (where there would be very little for them to eat) and are reported from the forested valleys in the foothills of these mountains across a vast swathe of Central Asia from the Caucasus mountains to Western China, involves a leap of faith and a paradigm shift that I, for one, simply cannot accept. The story of the yeti will continue yet awhile.

Slainte,

Jon Downes (Director, CFZ)

Newsfile

Enter the Olinguito

The genus *Bassaricyon* is a group of small procyonids, popularly known as olingos. The number of species in the genus Bassaricyon has been a matter of contention for some years.

Some taxonomists recognised as many as five separate species, and others recognised only two. They are quite rare in captivity and often misidentified as kinkajous.

An undescribed olingo similar to but distinct from *B. alleni* was discovered in 2006 by Kristofer Helgen at Las Maquinas in the Andes of Ecuador. *B. neblina* or the olinguito was finally described this summer.

Helden and his team using a multidisciplinary approach have now confirmed that four known species can be identified in the genus. These are:

- *Bassaricyon alleni,* lowlands of Guyana, Venezuela, and in Colombia, Ecuador, Peru and Bolivia east of the Andes;
- *Bassaricyon gabbii,* Central American,

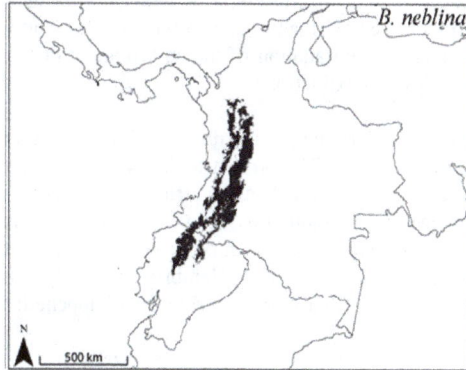

lowlands and highlands of Nicaragua, Costa Rica and western Panama;

- *Bassaricyon medius,* lowlands of Panama and in Colombia and Ecuador west of the Andes;
- *Bassaricyon neblina,* a montane species endemic to cloud forests in the Andes of Colombia and Ecuador.

The first specimens were actually discovered in museum collections in Chicago. Helgen's team wanted to clear up the question of how many Bassaricyon species there were once and for all, and examined more than 95% of the world's museum specimens of olingos. They were surprised to find a completely new species lurking unsuspected amongst them.

"The discovery of the olinguito shows us that the world is not yet completely explored, its most basic secrets not yet revealed," said Kristofer Helgen.

"If new carnivores can still be found, what other surprises await us? So many of the world's species are not yet known to science. Documenting them is the first step toward understanding the full richness and diversity of life on Earth."

A Painted Frog already

The rediscovery of the Israel painted frog, (*Latonia nigriventer*) is a story which has several interesting facets to it. It is oft used as a 'poster species' for those who wish to show how mankind's degradation of the environment drives species into extinction.

In 1951 work began on draining Lake Hula in northern Israel. The project took seven years and was initially hailed as a triumph of Israeli engineering, although as the years passed it was seen to have been a mixed blessing. Tamar Zohary and K. David Hambright writing on jewishvirtuallibrary.org describe what happened:

> With time, severe problems developed in the middle and southern parts of the Hula Valley, most of which originated from peat sediment degradation and subsidence. As the level of the groundwater table fell, air penetrated into the dried peat enhancing microbial decomposition of organic matter. Often these processes led to uncontrollable underground fires and the formation of dangerous caverns within the peat. The weathered peat soils turned into infertile black dust. Strong winds sweeping the valley produced dust storms that caused major damage to agricultural crops. Consequently, the ground surface subsided by up to three metres in some regions and inundation of these areas during winter rains restricted cultivation in many areas. An indirect problem associated with the drying of the soils was the proliferation of field mice populations which soared and wreaked havoc on agricultural crops in the valley. Over time, farmers abandoned more and more of the valley where cultivation was no longer profitable, thereby further enhancing the rate at which these soils deteriorated.

> In addition to these agricultural problems, various ecological problems became apparent. The decomposing peat released large amounts of nitrates

and sulfates which during the winter rainy season were washed into Lake Kinneret.

Although the impact of sulfates on the Kinneret is less obvious, nitrates are major sources of the nitrogen required for algal growth and their addition to water can lead to reduced water quality. Estimates are that about 40 per cent of all nitrate loading into Lake Kinneret comes from the drained Hula Valley

The Israeli government tried to reverse the

damage some years ago and re-flooded parts of Lake Hula but it was too late for two species of fish - a cyprinid and a cichlid - which had been driven to extinction, and in 1996, a third species - the Israel painted frog was declared to have become extinct, probably in the 1950s, and almost immediately became the aforementioned totem species of the environmental lobby.

However, to everyone's surprise and pleasure the species was rediscovered in 2011, and conservation measures were immediately brought into being. But the story is not over yet. Because it turned out that the Hula painted frog isn't actually a painted frog after all, but is actually something far stranger.

In an article for *Nature Communications* published in June 2013, Rebecca Biton *et al* describe how:

> Amphibian declines are seen as an indicator of the onset of a sixth mass extinction of life on earth. Because of a combination of factors such as habitat destruction, emerging pathogens and pollutants, over 156 amphibian species have not been seen for several decades, and 34 of these were listed as extinct by 2004.
>
> Here we report the rediscovery of the Hula painted frog, the first amphibian to have been declared extinct. We provide evidence that not only has this species survived undetected in its type locality for almost 60 years but also that it is a surviving member of an otherwise extinct genus of alytid frogs, *Latonia*, known only as fossils from Oligocene to Pleistocene in Europe. The survival of this living fossil is a striking example of resilience to severe habitat degradation during the past century by an amphibian.

Allegedly plans are afoot to flood more parts of the Hula Valley, and it is sincerely to be hoped that the vicious and internecine political squabbles in the region will not get in the way of this species finally being given a chance of a secure future.

Behold the bone skipper

Bone skipper flies were originally described by the seminal German naturalist Georg Wolfgang Franz Panzer in 1794. He gave them the Latin name *Musca cynophila*, and the German name Hundefliege ("dog-fly"), having found it on the carcass of a dog in Mannheim.

The peculiar thing about these flies is that they prey on the decaying carcasses of large animals, and for 160 years after it was last sighted in 1849 it was thought to have become extinct largely due to changes in livestock management in central Europe, improved carrion disposal following the Industrial Revolution, as well as the eradication of wolves and other big bone-crushing carnivores. "Because of that, *T. cynophila* was claimed as the first case of a fly species eradicated by man," says Dr Daniel Martín-Vega of the University of Alcalá, to the east of Madrid in Spain.

In 2010 it was rediscovered in Spain, and in 2012 our friend and colleague Lars Thomas received a report from a friend of his in Denmark:

> "He had been out one day taking a walk, when he stumbled upon a very dead something - probably a fox or a dog. It stank to high heaven, and as could be expected, it was absolutely covered in flies. Most of

these were blowflies of one species or another, and as such rather dull. But there was one that was rather special. It was fairly large, something like a centimeter, bluish black, none of which was remarkable in any way. But this thing also had a big round orange head! He described the fly as looking like it was ferrying a brightly coloured pearl around".

This sounds very much like a bone skipper. If we could understand how and why these macabre little creatures have returned after 160 years languishing in the zoological wilderness, I have a sneaking suspicion that we would know more about the workings of the universe than we do at the moment.

But it is not just a single species, lost for a century and a half that has re-appeared. According to Douglas Main writing on livescience.com in July 2013:

> In the past few years, three species of bone-skipper have been rediscovered in Europe, setting off a buzz among fly aficionados. But many bone-skippers were found by amateur scientists and recorded in photographs or video; actual specimens of the flies are few and far between. For the first time, Cerretti and colleagues have established a "type specimen" or "neotype" for one bone-skipper species, to which all of these bone-skippers will be compared in the future, in order to be identified.

Pierfilippo Cerretti, a researcher at the Sapienza University of Rome told how the fly's "previous taxonomy was almost completely incorrect - a mess. If you have no good specimens, you have no good taxonomy." He went on to explain how the newly typed species, *Centrophlebomyia anthropophaga*, was first described by a scientist in 1830 "based solely on his memory of specimens he had observed in large numbers destroying preparations of human muscles, ligaments and bones in the Paris School of Medicine in August 1821.

Man Beasts (BHM)

The Stoneman picture

Pennsylvania hiker John Stoneman claims to have taken the picture below, which many people have identified as being better than average shots of a mythical North American man-beast. The trouble is that they may well be nothing of the kind.

Stonehouse is quoted as saying:

> "I'm a skeptic myself. I'm not a believer, but this was not a bear and you can see fur on it ... It's wider at the shoulders and tapers down whereas a bear is bigger in the middle and stands differently with its paws out -- this was standing like a man, like a Bigfoot."

The story was given massive coverage online but

then the above picture was sent to *The Huffington Post*. Some weeks later an anonymous tipster who still - at the time of going to press - has not been identified, sent the above photograph to HP claiming that it shows nothing more than a tree root upturned by a mechanical digger. However, Stoneman has apparently refused to comment and the anonymous tipster has gone to ground.

The whole thing smells somewhat of fish methinks.

The Tasmania Expedition

At the time of going to press, an international expedition arranged by CFZ Australia has just returned to the UK. There will be a full write-up in the next issue, but we couldn't let such a historic occasion pass without asking Richard (who arrived back at the CFZ late last night) for a few words:

Tasmania is a very big 'little' island. About the size of Ireland, its population is less than a million, most of whom live in Hobart. Towns in Tasmania are more like large villages, and villages are like flea specks. There are masses of wilderness; deep, thick forests not only inside national parks, but in the countryside in general. There are massive tracks of wilderness in the north-east, north-west and the south-west. The latter, about the size of the West Country in England, has no human population, no roads and no habitations.

I was amazed at the amount of wildlife and the lack of humans. The roads are mostly overgrown and unused in the wilds of the north-west, with old logging tracks slowly being reclaimed by nature as they lie unused for years. Here I have seen more wildlife than anywhere else than Sub-Saharan Africa; pademelons, Bennett's wallabies, wombats, potoroos, echidnas, Tasmanian devils, tiger quolls, tiger snakes, wedge tailed eagles, boobok owls, native hens, sulphur crested cockatoos, black cockatoos, ground parrots and many others were seen.

We interviewed witnesses including a government licensed shooter who had seen thylacines twice in recent years along remote roads. The habitat is perfect for them and the prey is plentiful. The disturbance by human activity is minimal, and in the south-west would be non-existent. Taken with the sheer number of witnesses, I have no doubt the thylacine survives in Tasmania.

This has been the first of what will be a series of CFZ expeditions to the island; an ongoing investigation which will continue until we have proven the creature's existence.

Aquatic Monsters

A Dino Down Under

Magnetic Island off the coast of North Queensland was the epicentre of another global media monster hunt, if that is what you call it when a whole slew of newspapers around the world get excited over something that they decide should be dubbed 'The New Loch Ness Monster'.

These things happen with depressing regularity, but in this case the picture is better than usual and slightly reminiscent of the 1977 Sandra Mansi photograph taken at Lake Champlain in Vermont.

Island resident David "Crusty" Herron, who took this photo, said he was glad to capture the animal or object on film, and had started calling it Lost Nessie.

> "I looked out and saw this thing in the water and thought 's---, it's a Loch Ness monster'. There was this feeling of excitement on the beach, and all these people were pointing and talking about what it could be.

Someone said it looked like the Loch Ness monster, and said maybe Scotland had been too cold lately so it decided to come and visit Maggie."

The lovely Sharon Hill at Doubtful News said what we were mostly thinking, and as she was the one who said it first, I shall respectfully quote her words:

> "Eyewitnesses automatically went to the cultural touchstone closest in their minds – the Loch Ness monster. Even though that's never been found or even had a common, agreed-upon descriptions. What we do know is that no animal holds its head like this but dragon boats do...
>
> There was mention of increased tourism, of course. But, there was no mention of the thing moving at all. So, yep, this looks like a cut and dried... boat."

Sharon, m'dear. As usual, I agree with you entirely.

An Austrian Incident

According to Wikipedia, Traunsee is a lake in the Salzkammergut, Austria, located at 47°52′N 13°48′ E. Its surface is approximately 24.5 km² and its maximum depth is 191 metres. It is a popular tourist destination, and its attractions include Schloss Ort, a medieval castle. At the north end of the lake is Gmunden, at the south end is Ebensee. The lake is surrounded by mountains, including the Traunstein, and a number of other towns and villages surround the lake, including Altmünster and Traunkirchen. However, there is something that Wikipedia will *not* tell you: apparently – at least according to British tourist Matthew Townsend – it also has a monster.

He emailed us with this photograph. I have to admit that at first sight I was not at all impressed: it looks like a monstrous rubber duck. But he was polite and when I telephoned him he told me the story in his own words:

> "Well basically I was in Austria – my parents have a house out there – so I occasionally go out for ski-ing, and I took my three year old daughter out there just for a bit of a holiday and we're walking around some of the lakes and the mountains and I just happened to spot something just splashing around really. So I got a bit closer and thought that doesn't look like anything I've ever seen. I took a few photos with my camera and that's about it really".

I asked for more details, and he told me that the event took place around May 20th this year and that:

> "...from a distance it was splashing around and then it took me about thirty seconds to a minute to get down from the hill where I was to the actual part of the lake where I stood and then I saw it for about 10 – 15 seconds after that."

When he got back to his parents' house, he spoke about the incident to their neighbour Christoph with whom he had become friendly a few years before. He wrote to me:

"Firstly the local name is either Wasser-Drachen (which literally means "water Dragon") or Tatzelwurm which he translated as "Thigh-worm!" Apparently it is a bad omen to speak of the animal and if someone does, it comes in the night to steal their children. It is also used to "scare" naughty children into being nice claiming that the "Tatzlewurm" will eat them if they are naughty.

Christoph mentioned that these days most people have either forgotten about it or do not speak about it due to the "bad omen" associated with it yet there are still quite a lot of sightings in different areas of Austria each year. He has apparently seen it as a child in the woods but it was such a long time ago he could not provide any other details aside from "It was f-ing huge!" which isn't that helpful!"

As any cryptozoologist will know the name 'tatzelwurm' is more usually associated with a stumpy reptile and amphibian reported throughout the Alps, which has variously been identified as an unknown salamander, skink or snake. Sometimes it is reported as having two limbs, sometimes four, but I have never heard of it being described as a fairly 'conventional' lake monster before.

My gut feeling, as with anyone else I know who has seen the photograph is that it is a hoax of some kind. But who is the hoaxer and who is the hoaxee? Matthew Townsend has always been helpful and polite during my dealings with him, and I have to say that in my opinion he is being entirely truthful about what he says he saw. So is someone trying to hoax him? If so, how and why? Or is my judgement of human nature woefully inadequate. Matthew tells me that he will be going back to Austria next summer and will try to find out more, so watch this space!

Ness than Convincing

One of the most convincing Loch Ness photographs of recent years has turned out to be a hoax.

At the time I wrote that it was one of the most convincing pictures for years. Well it bloody well should have been! It turned out that George Edwards, the skipper of one of the most popular tourist boats on the loch had simply photographed a fibreglass hump that had been commissioned by

National Geographic for a 2011 documentary.

Mr Edwards had been involved in the making of the documentary, and - as he had access to the props and equipment - simply took the 'hump' out into the deep water and photographed it. Steve Feltham, who has spent much of his life on the shores of the Loch became righteously angry, and is quoted as saying.

'It does the subject no good and damages his own reputation. When you read things like this in the papers, people will think it's all just a fairytale. But if you read the reports and books you're more likely to think that something is there to be explained. He's supposed to be taking people out on tours but he's nothing more than a faker and a liar.'

However, I think that George Edwards' words may actually give a better insight into the true feelings of the residents of the area, who - I suspect - are heartily sick of the subject.

'Where would Loch Ness be without the world's best known forgery, the Surgeon's Photograph? These so-called experts come along with their theories about big waves and big fish, and their visitor centre, but I'm sick to death of them. People come here for a holiday and a bit of fun. I'm one of the people who has brought thousands of people to the Highlands over the years, and I can tell you they don't come here for the science.'

Carl Marshall's Column

In 2010 I travelled, along with my colleague Andrew Jackson (Geordie) from the Stratford upon Avon Butterfly Farm, to Central America, into the jungles of Belize, to study the country's wide diversity of ecosystems and to record the species we found within these habitats. In many ways the expedition was a success. We found many interesting invertebrates; some of which we had never seen before, we heard and then later saw howler monkeys, and the highlight of the trip was when we witnessed and photographed a sleeping jaguar high in the tree branches on the outskirts of a reserve. Diverse as the wildlife in Belize was, as we were eating spiced noodles and drinking coffee one evening around the campfire,

we discussed the possibility of an expedition to Borneo in Southeast Asia to experience the island's greater diversity of larger fauna. Borneo had always been one of my biggest aspirations, mainly because I knew it would test my endurance capabilities; this plus the strong desire to see wild forest elephants, gibbons, proboscis monkeys and of course the man of the forests - the orang utans in their natural environment.

Borneo's rainforests have been dated at 130 million years old, making them some of the oldest rainforests in the world. There are said to be 15,000 species of flowering plants with 3,000 species of trees of which 267 are of the titanic Dipterocarpaceae family, one species Shorea fageutiana reaching 88m - the same height as the tower at Westminster Cathedral give or take about a foot! Borneo has 221 species of terrestrial mammals and 420 species of residential birds. There are approximately 440 freshwater fish species, which is about the same as Sumatra and Java combined; 149 of them are endemic to Borneo. Borneo is an important refuge for many forest species, including the Asian elephant; the Sumatran rhinoceros, the Borneo clouded leopard, the Hose's civet and the Dayak fruit bat.

The Sumatran rhino is one of the most endangered animal species in the world due to hunting for its horn and deforestation over the past few centuries. Today its population has dramatically decreased. At the beginning of the 20th Century the Borneo sub-species of the Sumatran rhinoceros was widespread over the island, but by the 1980's loss of habitat through conversion to permanent agriculture - particularly palm oil plantations - had become a significant threat. The Borneo Sumatran rhino is now probably extinct in Sarawak and were believed to be extinct in Kalimantan until rhinoceros footprints and evidence of foraging were reported in west Kutai, East Kalimantan by the World Wildlife Federation (WWF) while monitoring orang utans earlier this year. With perhaps fewer than 200 individuals surviving in Indonesia and Malaysia, this prehistoric-looking rhinoceros must be protected at all costs. Only two areas in Sabah contain rhino populations which have good

Borneo: The Jackson-Marshall expedition part 1

© 2013 Carl Marshall

prospects of long-term survival with adequate protection and management. The Tabin population was under pressure from forest loss and was afforded protection by the Sabah government in 1984 through the establishment of the Tabin Wildlife Reserve. The Ulu Segama/Kuamut population is scattered through a vast area of several contiguous forest reserves including the Jurassic looking Danum Valley.

In March 2013 Geordie and I travelled into the interior of Malaysian Borneo to study its ecology and biodiversity and were lucky enough to encounter many wonderful vertebrate species such as the forest pygmy elephants (not as small as the name suggests), a slow loris, the pig-tailed macaques, crocodiles and many unique species of bats.

On this trip Geordie and I were not just looking for any evidence of rare known fauna, we were also inquiring after any cryptozoological reports. Richard Freeman, the zoological director of the Centre for Fortean Zoology (CFZ), had advised me to be aware of reports pertaining to the Bornean

orang pendek - the batatut, as very little research has been carried out to identify the mystery surrounding this creature and we were charged to collate any information from any sources that contributes to demythologizing this enigma. The list below contains some hitherto undocumented cryptids...

- Giant saltwater crocodile: A soldier informed us of a colossal, 35ft crocodile he had witnessed near Lok Batik, Sabah.

- Giant reticulated pythons: Our guide from Ulu Kamanis informed us of giant pythons 30ft+. More research on these in 2014 expedition.

- Luminous birds of paradise: We were told of strange glowing paradise type birds in the deep forests of Ulu Kamanis.

- Giant black orangutan: Flying Snake editor Richard Muirhead had previously informed me of oversized orang utans being reported from Borneo. Upon inquiry these seem to come from Indonesian Borneo and often appear to be uniform black in colour.

© 2013 Carl Marshall

- Sabah sky rods: Matthew Lazenby has researched thoroughly the flying rod phenomenon. He has taken part in a reconnaissance of a deep cave, that after viewing back the video footage shows very strange flying rod shaped objects that Matthew claims display a different flight pattern to those of moths filmed using extremely short exposure times. Moths or rods - we will meet with Matthew (Jigger) again in 2014 and will hopefully have further updates on this persistent phenomena.

- Batatut: The Bornean answer to orang pendek is named the batutut. I found no credible evidence of this cryptid from Malaysian Borneo! Further research on this mystery hominoid in 2014 expedition into Indonesian Borneo.

- Possible out of place (oop) Huntsman spiders: While staying at Tampat Du Aman we briefly witnessed a Heteropoda sp that very closely resembled H. davidbowie, a species supposedly only found in peninsular Malaysia - could this species also be found in Borneo? We will investigate this possibility further in 2014.

- Oop long arm scarab beetles: Max Blake of the CFZ informed me last year that because of the distribution of many species of Euchirinae, specimens might be found in the middle east and incredibly on the island of Borneo. No evidence of these currently! More research in 2014.

- Horned cat: An alleged horned felid from Indonesian Borneo. More in 2014.

This list will be updated in 2014 when we return to Borneo for our followup expedition - next time mainly exploring Indonesian Borneo. Watch this space!

Carl Marshall works at Stratford Butterfly Farm and is a fine field naturalist. Over the past couple of years he has become a very enthusiastic member of the CFZ, and his quasi-fortean view of British natural history fits in perfectly with my own. He was, therefore, the perfect choice as a columnist for the brave new *Animals & Men*, and we are proud to have him aboard.

I was bumbling around one evening in one of the Costa Rican rainforests on the Osa Peninsular in January 2013. I was actually on the hunt for two creatures, specifically the infamous Fer-de-Lance snake (*Bothrops asper*) and the quite remarkable red-eyed tree frog (*Agalychnis callidryas*). I knew the frogs lived high in the trees but descended very occasionally to a small pond. The asp also knew this and sure enough one was waiting coiled and ready for a strike about a metre away from me. As always on these exhilarating night walks there were a good many distractions to keep me preoccupied and I also found spiders, scorpions and a very odd looking Amblypigid.

However, such a curious creature paled into insignificance in comparison to what I was about to stumble upon. Affixed to the stem of a plant at the side of a small pond was a brown twig – at least I thought it was a twig but then the guide I was with asked 'what do you think that is? *(see figure 1)*. I proffered that it was a twig but no it was in fact a bagworm moth, also known as a case moth. They are quite bizarre invertebrates

and even now I know very little about them so I will have to go back one day for a period of serious study in the field.

These 'Psychidae' are a group of small moths that are members of the Superfamily Tineiodea. Female adults of most species lack wings or have very small and non-functional wings. Males are typically black and their wingspan is 1.2-3.6 cm. The abdomen of these moths is long and tapering. Adults of both sexes have vestigial mouth parts. In some species, females entirely lack eyes, antennae, and legs. Larvae form characteristic spindle-shaped silken cases covered with bits of leaves, twigs, and other debris. Each larva enlarges its case as it grows. These moths pupate

animals camouflage themselves in rainforests and these specimens were utterly remarkable. They blend perfectly into their surroundings and I still do not know what if anything might locate and eat it at this stage. I cannot help but muse, how did such a creature come about? When I find something like this I am given to deep cogitation. Is this is a product of evolution or is it simply here by design? If by design then who or what brilliant being could possibly have thought of such a creature? If by evolution then how much has it changed/adapted and over what period of time?

Apparently even some of the locals have never seen any adults of these wonderful moths on the Peninsula and at least one person I spoke to would not accept that they existed, yet exist they must and I will make it my business to find one. After all, that's where curiosity can take you…

in the larval case after it is attached to a twig with silk. In most species, the female does not leave the case, but attracts males by emitting pheromones from her abdomen. To mate, the male thrusts his abdomen through the open lower end of the case. The female lays her eggs in the case; when they hatch, larvae crawl away to feed and form their own silken cases. How fascinating!

After I saw this I managed to find another example, this time on my own – *see figure 2.* Now I hope you are taking notes as I will be asking questions afterwards. There are eleven known subfamilies and around 240 genera among the bagworms. They can be found globally. I wondered what an adult looked like since none were seen in Costa Rica during my time there. See *figure three* which is simply an example I took from the Internet.

What fascinates me is that some bagworm species are parthenogenic, that is to say they do not require male fertilisation to develop the eggs. Further, I have always been interested in the way

Born in Birmingham, England Carl Portman has always followed the maxim 'interest is where you find it' and this certainly applies to natural history. He has bred endangered species of tarantula spiders, written two books on natural history travel and lectures around middle England on animals and rainforests. Oddly he has a diploma in sexing juvenile theraphosid spiders, is an English Chess Federation County Chess Master, supports Aston Villa and has a strange addiction to Turkish Delight (covered in chocolate). Having worked for the Ministry of Defence for 30 years he now spends his time doing lecturing, chess coaching, some photography and management consulting.

He has spent time studying animals in the rainforests of Australia, Ecuador and Costa Rica searching for new and ever curious insects and arachnids and has a desire to find a new species somewhere in the world. His motto is 'Don't complain about the dark, light a few candles'.

He is married to Susan and lives in Oxfordshire. Their two Border Collies, Darwin and Dickens keep them fit and ensure that there is never a dull moment in the household.

In 2012, Anomalist Books published what I personally consider to have been the finest cryptozoological title of that year: Lyle Blackburn's *The Beast of Boggy Creek*. It quite rightly received excellent reviews and justifiably thrust both Lyle and the Fouke Monster, itself, firmly into the limelight. Sometimes, however, when an author puts out such a great *first* book it begs an awkward question: how do you follow it with an equally good *second*?

Of course, Lyle could have done the literary equivalent of what the Sex Pistols did with their classic 1977 album, *Never Mind the Bollocks*. That's to say, the band put out an amazing record and then split only months later. Their legacy: a still-stunning, stand-alone collection of anthems from a group that went out with a bang.

Fortunately for us, the readers and fans of all things monstrous, Lyle chose not to follow the "live fast, die young" approach of the Brit punks. Instead, just about as soon as he was done with ol' Boggy, Lyle was hot on the trail of the truth surrounding yet another fantastic creature: Lizard Man.

As is very much also the case with such famous cryptids as Mothman and Owlman, its name provokes imagery of a comic-book-style super-villain. But, in reality, and as Lyle demonstrates, there's nothing cartoonish about Lizard Man in the slightest. "Nightmarish" is a far better description.

So, with that said, what has Lyle served up for us this time?

The Lizard Man: The True Story of the Bishopville Monster is an enthralling, chillingly atmospheric, and deeply revealing look at a strange and controversial legend that first surfaced in 1988. This is a story of small-town secrets, of a cast of fascinating and diverse characters, of media hysteria, and of a terrible, and terrifying, animal that just may be more than the myth that many assume it to be.

It's important to note, too, that even though the story is more than a quarter of a century old, its fascination continues to endure – within Bishopville itself, within the mainstream media, and within the field of cryptozoology, as Lyle skillfully demonstrates.

What I particularly enjoyed about *The Beast of*

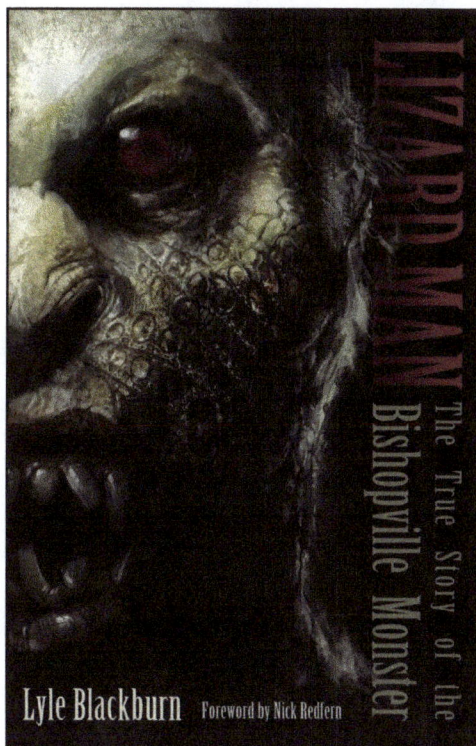

LIZARD MAN
The True Story of the Bishopville Monster
Lyle Blackburn   Foreword by Nick Redfern

Boggy Creek is that Lyle didn't lazily sit at home, while securing all of his data from the Internet. Instead, he hit the road and sought out the truth, firsthand. And that's precisely what Lyle did while researching the story you will very soon be reading.

With his research partner, Cindy Lee, Lyle headed off - road-trip style - to South Carolina, where he tracked down the key players in the saga (or, at least, those who have not fallen foul of what may be nothing less than a sinister "Lizard Man Curse"), and he explored the wild woods and the spooky swamps of Bishopville, with just one thing on his mind: finding the answers to the beastly puzzle.

It wasn't as easy as it might sound, however. Not only have 25 years passed since the Lizard Man first surfaced, but, as Lyle learned and as he demonstrates to us, a lot of misconceptions about the critter, the reports, the eyewitnesses and much more besides have been made.

Rather like Sherlock Holmes and Dr. Watson roaming the foggy wilds of Dartmoor, England, in pursuit of the Hound of the Baskervilles, Lyle and Cindy undertake far more than a bit of dicey detective work as they make their way around town and the surrounding, woody, swampy landscape. And also just like Holmes and Watson, our dynamic duo uncover a wealth of untapped data and cases, they encounter a variety of people who may be the key to certain aspects of the saga, and they realize that Bishopville really does appear to be the domain of something wild, something primitive.

Packed with cool artwork and photos, and written in a highly entertaining and atmospheric style that pulls the reader right into the black heart of the mystery, *The Lizard Man* is an important piece of work. And it will justifiably remain so, too.

Now, Lyle, what about book number three…?

Nicholas "Nick" Redfern, born 1964 in Pelsall, Walsall, Staffordshire, is a British best-selling author, Ufologist and Cryptozoologist now living in Dallas, Texas, U.S.

Redfern is an active advocate of official government disclosure of UFO information, and has worked to uncover thousands of pages of previously classified Royal Air Force, Air Ministry and Ministry of Defence files on UFOs dating from the Second World War from the Public Record Office and currently works as a feature writer and contributing editor for Phenomena magazine.

Nicholas "Nick" Redfern, born 1964 in Pelsall, Walsall, Staffordshire, is a British best-selling author, Ufologist and Cryptozoologist now living in Dallas, Texas, U.S.

Redfern is an active advocate of official government disclosure of UFO information, and has worked to uncover thousands of pages of previously classified Royal Air Force, Air Ministry and Ministry of Defence files on UFOs dating from the Second World War from the Public Record Office and currently works as a feature writer and contributing editor for Phenomena magazine.

watcher of the skies

CORINNA DOWNES

- **Little bitterns (*Ixobrychus minutus*) breeding in Somerset**

Little bitterns have bred again in the UK. The RSPB Ham Wall nature reserve in Somerset announced that guards had been set up to protect these rare birds – the only known breeding pair of this member of the heron family in the UK - and their nest. They first nested in Yorkshire in 1984, but not again until 2010, again at Ham Wall. Although breeding has been suspected since then, this year they were known to be nesting, and the young successfully fledged. The little bittern is in decline across Europe presumably due to loss of habitat, both where they nest and in Africa, where they spend their winter.

Source: http://www.birdguides.com/home/default.asp
Image: Ferran Pestaña/Wikimedia Commons

- **White-throated needletail (*Hirundapus caudacutus*)**

A white-throated needletail – a species not seen on our shores for 22 years – turned up at Tarbert on the Isle of Harris in the Western Isles in June. Dubbed 'bird of the century', its long-awaited return ended in tragedy, however, when it collided with the only wind turbine in the south of the island and was killed. This bird is the world's fastest-flying and breeds in eastern and southern Asia and winters in

south east Asia and Australia. It lives in forest and open country, feeding on insects during flight. It was first recorded here in 1845 in Essex and is also known as the spine-tailed swift and the white-tailed swift. A very sad ending indeed for this small master of flight.

Source: http://www.birdguides.com/home/default.asp
Image: Aviceda /Wikimedia Commons

- **Acsension frigatebird (*Fregata aquila*)**

First seen on 5th of July at Bowmore on Islay, and classed as an accidental visitor, an Acsension Frigatebird caused the sensation of the month, if not the year.

This bird (as the name suggests) breeds on Ascension Island, in the tropical Atlantic Ocean, an isolated volcanic island situated almost midway between the African and South American coasts. More specifically, the bird breeds on a single outlying island (known as 'Botswain Bird Island') but after the extermination of feral cats and rats, the

bird has recently recolonised the main island. The bird is pelagic; its range at sea being to the west African coast. The only other British record of Ascension Frigatebird was a juvenile found moribund on the 9th July 1953 on Tiree in Argyll. This bird feeds mostly on fish, but also squid, and jellyfish. It feeds from the air and will often harass other birds to steal their food.

Source: http://www.birdguides.com/home/default.asp
Image: Drew Avery /Wikimedia Commons

- **Cape May Warbler (*Setophaga tigrina*)**
On the 23rd October a Cape May Warbler was seen at Baltasound, Unst. The only previous record was of a male in Paisley in Clyde, Renfrewshire in June 1977. It breeds across the northern United States and the boreal forest of Canada and winters in the West Indies.

The species is named after Cape May in New Jersey, where Alexander Wilson collected the first type specimen, and somewhat eerily remarkable, Wilson's birthplace was Paisley, Clyde, his home being visible from the very trees in which Britain's first vagrant Cape May Warbler was found singing.

It feeds on insects, especially spruce budworms, during the breeding season, and nectar and insects in the winter. It has a unique tongue among warblers, whereby it is curled and semitubular and is used to collect nectar during the winter.

Source: http://www.birdguides.com/home/default.asp
Image: Peter Wallack

- **Mourning Dove (*Zenaida macroura*)**
On the 28th October a mega rare Mourning Dove was seen on the Isle of Rhum, Highland. Despite being just the third record for Britain, after the last individual in 2007 on North Uist, Western Isles and the first in 1999. It lives in open forest and farmland and feeds on seeds, especially those of cereals.

Once they've filled their crop - the record is 17,200 bluegrass seeds in a single crop! - these birds can fly to a safe perch to digest the meal. They sometimes eat snails.

Mourning Doves eat roughly 12 to 20 percent of their body weight per day, or 71 calories on average. Its range encompasses North and Central America, and is the most widespread and abundant game bird in North America. Every year hunters harvest more than 20 million, but it remains one of their most abundant birds with a U.S. population estimated at 350 million. The oldest known Mourning Dove was a whopping 31 years 4 months old.

Source: http://www.birdguides.com/home/default.asp

- **'Lost' bird rediscovered in New Caledonia along with 16 potentially new species**

In early 2011, Conservation International (CI) dubbed the forests of New Caledonia the second-

most imperilled in the world after those on mainland Southeast Asia. Today, CI has released the results of a biodiversity survey under the group's Rapid Assessment Program (RAP) to New Caledonia's tallest mountain, Mount Panié.

During the survey researchers rediscovered the 'lost' crow honeyeater (*Gymnomyza aubryana*) and possibly sixteen new or recently-described species.

Over 20 percent larger than Connecticut, New Caledonia is a French island east of Australia in the Pacific Ocean.

Source:
http://news.mongabay.com/2013/1029-hance-rap-new-caledonia.html#BKjOyO6YK78tBk6z.99
Photo: Wikipedia/ Richard Fuller

- **Owl recorded in Oman could be a new species**

Ornithologists working in Oman say an owl

discovered in a remote, mountainous region could be a new species.

After repeated trips to the remote site, Magnus Robb and naturalist and photographer Arnoud van den Berg captured photographs of the bird.

"I was listening through my headphones, when I suddenly heard something completely different [to the owl species I was there to record]," he told BBC News. "I know the other Arabian owl sounds quite well, and this was clearly something that didn't fit."

Source: http://www.bbc.co.uk/news/24374313

- **Fifteen New Species of Amazonian Birds**

In August an international team of researchers coordinated by ornithologist Bret Whitney of the LSU Museum of Natural Science (LSUMNS), recently published 15 species of birds previously unknown to science. Not since 1871 have so many new species of birds been introduced under a single cover.

"Birds are, far and away, the best-known group of vertebrates, so describing a large number of uncataloged species of birds in this day and age is unexpected, to say the least," said Whitney.

"But what's so exciting about this presentation of 15 new species from the Amazon all at once is, first, highlighting how little we really know about species diversity in Amazonia, and second, showing how technological advances have given us new toolsets for discovering and comparing naturally occurring, cohesive ('monophyletic') populations with other, closely related populations."

Source:http://www.sciencedaily.com/
releases/2013/08/130828173115.htm

Orang Pendek: The Evidence Mounts

Richard Freeman

About half-a-dozen times a year I get contacted by TV companies saying that they want to make a documentary about cryptozoology. Mostly they come to nothing. Once in a blue moon one actually amounts to something. Late in 2012 I was contacted by a French film maker called Christophe Kilian. Chris had worked on a number of projects for Scienti Films, a French company that specialises in science documentaries.

Chris wanted to make a film about 'wildmen' around the world but focusing mainly on orang-pendek, the upright walking, powerfully built, but relatively short, mystery ape of Sumatra. We exchanged a number of e-mails and information. Some information was completely unknown to me. Chris told me of several alleged yeti hairs held by the Natural History Museum in Paris. The hairs were taken from a stick that a mountaineer used to hit the back of a yeti. Some hairs stuck to the stick and were brought back to France. One was examined and pronounced to be similar to that of an orang-utan. Unfortunately, like most relics of cryptozoology, they have now been lost.

Chris had read my book *Orang-pendek: Sumatra's Forgotten Ape* and wanted to return to the jungles of Kerinci Sablat National Park with me to look for evidence of the creature.

The project was pencilled in for January but was delayed until late June/early July, which, as it turned out, was a piece of serendipity. Joining us would be talented artist and taxidermist Adele Morse. I had met Adele in 2011 when she contacted me wanting information on orang-pendek. Adele had found my writings about the creature online and had decided to do an installation, or art project, on the creature for the Royal Academy. I gave her as much information as I could, and sent her a copy of my book.

Adele wanted to interview and film me in person. She came down to Exeter from London for the weekend. Adele, originally from rural Wales is a self-taught taxidermist who had been stuffing animals and creating unusual sculptures for years. She was best known for a fox she had stuffed many years before. The cross-legged vulpine had staring, slightly cock-eyes that made it look like it was stoned, or perhaps mad. She was going to throw it out but decided to put it on E-bay. It sold for £300! The buyer put some pictures of it on line. In a truly Fortean set of events the fox became an internet meme in Russia. It was Photoshopped with just about everyone from Stalin to The Simpsons. So popular was the fox that Adele took it on two tours of Russia, all expenses paid, and was greeted by a throng of over 100 journalists at the airport. There has even been a bronze statue of it erected in a remote Russian town near the border with China.

Adele's installation consisted of a number of life-sized orang-pendeks and a shop selling orang-pendek related merchandise that was based on a Sumatran hut. I visited the opening with my old mate and orang-pendek witness Dave Archer. Adele's project came second in a national competition.

Adele and I travelled from Heathrow and after a 22 hour flight via Jakarta we touched down in Padang. We were met by the fixers and began the long drive from Padang down to Polempek, the village of my former guide Sahar Dimus.

Do to road maintenance on the mountain route we took the more scenic coastal road. Stopping for dinner in a seaside restaurant I, for the first time, tried fresh durian. Despite visiting Sumatra four times previously I had never had the dubious pleasure of tasting this infamous fruit. Depending on who you ask, the flavour is disgusting or heavenly, having been described as like a rotting quiche or like custard and almonds. The odour is, however, universally derided and has caused many Asian hotels to ban the fruit.

The spiky outer husk was cut away with a parang and inside were the yellowish-white, fleshy, edible masses surrounding the seeds. Each segment was about the size of a lemon. The smell was a mixture of stale urine and rotting onions. Undeterred I bit into the pulp. It had a consistency like thick mousse. The taste was akin to rotting liver pate. I managed two pieces. Adele and Chris could only stomach one. The fixers happily scoffed the rest with relish.

Paddling in the sea outside of the restaurant Adele came upon a tiny octopus but the little cephalopod was too swift to photograph.

Later we stopped by an old shack inhabited by a troop of long tailed macaques. We filmed and photographed them after attracting them with grapes, bread and bananas.

Sahar's village, Polempek, lies on the outskirts of Kerinci Sebelt National Park a hot spot for orang-pendek sightings and the focus area of our past expeditions. Sahar had been my guide on the four previous expeditions I have been on in Sumatra. He had lived for fourteen years on the outskirts of the jungle but only saw the orang-pendek once, with Dave Archer in 2009. Sahar died of kidney problems just weeks after our 2011 trip. We had no idea that he was ill and he was only 41 when he passed away, leaving a wife and three sons. The picture opposite was drawn by him.

The CFZ ran a collection to send Sahar's widow, Lucy, money to care for her children. Sahar's brother John also stepped in to help. We visited Lucy in her house and found that she was doing well. John and Lucy's eldest son Raffles were now both guides themselves and would be guiding us on our track.

The fixers had arranged for a group of orang-pendek witnesses to meet us in Polempek. All of the men were locals and had seen the creature or its tracks in the area within the last year. With the fixers translating, Adele and I interviewed the men whilst Chris filmed us.

Herman Dani had seen the creature Uhan Danda a year ago. He only got a good look at the head. The face had a flat nose and thin eyes. Its fur was grey. The creature stared at him and he ran away.

Salim had seen strange tracks in the jungle five months before. He said they were man-like and the size of a human hand.

Amri had his encounter at Padutingi, about four hours from Polempek, seven months previously. The creature he saw was one metre tall with grey fur. He ran away in fear.

Juha Rapti had come across prints 8 months before at Sungi Minya, around four-and-a-half hours away. They were human sized but had a highly separated big toe.

Rahman saw the orang-pendek five months past at Gunung Sanka about one hour away. The creature was large with black hair that faded to grey. It was moving quickly and he didn't get a good look at it. He fled.

Saba Rudin saw the creature as it crossed a jungle trail. The area was between Sungi Mina and Sungi Kuni; about five hours away. The event had happened ten months before. The orang-pendek had a broad, barrel-like chest and black and grey hair. It walked on two legs like a man.

Aprisal was in the Sungi Kuni area nine months ago hunting wild pig. When he paused he saw a creature with black and grey fur, and a large mouth. Afraid he ran away. The area was about four hours distant.

Mah Darpin saw the orang-pendek after rainfall at Gunnung Kacho, nine hours distant. He saw the creature from the back noting that it had long fur of a grey-black colour, was around a metre tall and walked on two legs. He became afraid and walked back the way he had come.

Saimi Alwi saw the creature one year before at Sungi Minya which is about four hours from Polempek. Whilst tracking he heard a noise and saw a creature squatting to eat kitan fruit. The animal was barrel-chested and muscular. It had black and grey fur and stood a metre tall.

When questioned, none of the witnesses knew about the orang-kardill, a second, more human-like creature said to live in the jungle, or the chigau a supposed gold furred, savage big cat. Their stories were remarkably consistent and I was struck by the lack of exaggeration.

The next day we began the arduous climb up the rim of the extinct volcano that forms Gunung Tuju, the lake of seven peaks. The jungles surrounding the lake are a hotspot for orang-pendek sightings. The first time I climbed up the crater back in 2003 I thought I was going to have a heart attack. On my second climb and all the subsequent ones I had learned to take my time and pace myself, making the climb much easier.

However, this time around we were late starting the climb and we found ourselves tracking under the noon sun. Adele suffered at first but rallied later one. Conversely I was ok at first but began to suffer badly later in the day.

The climb consists of a steep, slippery, winding path that rises like a set of never-ending steps up to

through the jungle to the rim of the crater. I was soon gasping and sweating. I took little rests, but still the ascent was hellish.

After resting at the top we descended to Gunung Tuju itself. I was horrified to see that the boulders at the edge of the lake had been defaced by graffiti. It seems that true wilderness is getting harder and harder to find.

My old friends, the wretched canoes, were waiting for us. As the evening was drawing in we decided only to cross half-way before making camp next to a fisherman's shack.

The following day we made our way via canoe to the small bay where we usually make camp. On the way I was horrified to see a group of twenty or so students tracking through the jungle, shouting and making an awful racket. I was worried that they would scare away any wildlife for miles around. To my horror they seemed to be headed for a small inlet next to the bay where we wanted to camp. Luckily they were not staying overnight, and soon went back the way they had come. I still thought that there was now little chance of seeing any wildlife.

We made camp. Adele, Chris and I had our own tents whilst the guides and fixers erected a pondok, a large shelter made from branches, palm leaves and plastic sheeting. As the sun was setting and a fire was being started a primate call unlike any other I have heard rang out from the forest. I am familiar with all the calls of the wildlife in this area. I have also worked with all these primates in captivity. This was the call of a primate, but one I had never heard before. The vocalisation went 'ho...ho...ho...ho'.

The guides and porters froze and John Dimus said 'orang-pendek'. Chris frantically tried to record the sound but it had ended by the time he had his equipment ready, and the call was not repeated. It had sounded relatively close to camp. An exciting start to the expedition.

In the morning we trekked into the jungle. There was much evidence of tiger activity. Within a mile of the camp we found tiger claw marks on a tree trunk. We came across two tiger kills. Both were Malayan tapir and the bones had been picked clean.

Adele took a coccyx and vertebra as well as some teeth for her art display.

It was late June and a fruit called the kitan by locals was ripe and falling. About the size of a pineapple with a reddish-brown, hard outer skin and a yellowish pulp inside, it is said to be favoured by the orang-pendek. I have not yet been able to find the scientific or western name for the fruit. The guides said that only the orang-pendek favours it. We found several that looked as if they had been chewed by something with human-like teeth. Close by John found a hair, but touched it with his fingers! Still we preserved it in ethanol.

We set up a camera trap and baited it with fresh mango.

The following day we decided to explore the far side of the lake and set out in the rickety canoes. The terrain on the far side is much steeper and the jungle seems damper. We climbed up a ridge and listened to siamang gibbons calling. Whilst the rest of us were listening to the gibbons Adele came across a track. It was on a slight slope, but it clearly showed the human-like heel and broader front foot typical of an orang-pendek.

I had brought some dental cement with me. This is a kind of fine plaster of Paris ideal for casting tracks. I had tested the substance out in my garden back in Exeter to see how much was needed to make a cast and to practice until I got the consistency correct. As the substance was heavy I carried only as much as I needed to cast a couple of tracks.

Adele offered to make the cast as she was a sculptress. However, I had not reckoned with the greater moisture in the rainforest and the more porous consistency of the ground. The casting of this one track took up all the dental cement that I had brought.

Adele tried to strengthen the cast with strips of cloth. Once dry we gingerly eased it from the damp earth. It was not the best track I have ever seen but the heel and toes were visible. The detail was somewhat blurred by the action of the bunching up of the earth around it on the slope. The cast cracked in two but Adele offered to take it back to London and repair it.

A few minutes later we came across two hand prints that looked very like the one found by Andrew Sanderson on our 2011 expedition. They had a palm roughly the size of a human palm but the fingers were much thicker, and sausage-like in shape. The thumb was smaller than a human's but bigger than an orang utan's. All in all the hand shape is more like that of a small gorilla.

I was gutted that I had run out of dental cement. On future expeditions I will know to take far more than I think I need.

On our return journey the wind cut up rough and the waves on the lake were threatening to swamp the canoes. We pulled into shore and lashed them together with vines and creepers. We paddled close in to shore so the return trip took us longer. It was dark by the time we reached camp again.

The following day, we followed another trail in the jungle in a different direction. Half-a-mile from the camp we came upon another orang-pendek track close to a rotting log. We also found and collected hair samples in the area. Whilst taking the samples we heard the call again. 'Ho...ho...ho...ho'. Once more Chris was foiled in his attempt to record the vocalisation.

We moved further into the jungle and found another rotting log. Beside it was a set of the most perfect orang-pendek prints I have ever seen. No more than a day old they were perfectly preserved in the damp soil. They clearly showed the long, man-like heel, four toes at the front and the offset big toe at the side. They would have made amazingly detailed casts. Once more I cursed at not having brought more dental cement. I had to make do with filming and photographing them.

I believe that there are two keys to finding orang-pendek in Kerinci Seblat. Firstly to visit when the kitan fruit is ripe in late June/early July. Secondly, rotting logs. There have been a number of reports of orang-pendeks seen ripping rotting logs apart to find grubs and insect larvae. We found the prints of at least three individuals around rotting logs.

In the same area we found many hair samples, far more than on any other expedition. I filled all my specimen jars and preserved the samples in ethanol. Again we heard the now familiar call 'ho...ho...ho...ho' but further away in the jungle depths. We found chewed kitan fruit in the area as well.

On the way back to camp we were confronted by a mass of vegetation, a labyrinth of branches made up of both dead trees and fallen trees that were still alive. It was a hard climb across, round and through the jungle maze. Adele did particularly well as she has poor depth perception due to her lacking certain eye muscles.

The following day we set out to collect the camera trap. On the way we found more tracks, but these not so clear as the ones near the rotting log. At the camera trap area we saw that the mangoes had remained untouched. On our way back we faced a second vegetation labyrinth like the one from the day before.

Adele found some more hairs which we preserved.

Back in camp we looked at the shots from the camera trap. It showed nothing except us putting it up and taking it down. I was not surprised. When testing the traps back in England we found that they had to be left up for weeks on end before they captured anything at all.

That night it rained hard. We had been lucky with the weather so far. On past expeditions torrential rain was a bane. We sat up in the pondok with the guides and porters as they did conjuring tricks. Looking out into the dark I felt the palpable feeling of being watched. I saw or heard nothing but I had a conviction that *something* was watching us from the jungle night and was very close.

Next morning, as we were breaking camp I found the footprint of a tiger only thirty feet away from the camp area.

We made the long crossing back over the lake and climbed down the side of the crater. The crossing and trek took most of the day. We saw monkeys and pig hunters as we reached the lower forests. At Polempek the cars were waiting to take us back to Sungi Penuh.

We checked into a hotel which was a nice change after being under canvas. We met up with Dr Achmad Yanuar of the National University of Java. A primatologist, Dr Yanuar had worked with Debbie Martyr and Jeremy Holden during the Flora and Fauna International orang-pendek hunt in the 1990s.

A short, genial, bespectacled man, Dr Yanuar unfurled a map of Sumatra and showed us all of the places he had hunted for the orang-pendek. He had investigated sightings and findings of tracks over most of Sumatra except the north. In all of his years searching he never saw the creature himself or found any tracks, but he interviewed many witnesses. It seemed that even twenty years ago reports of orang-pendek were much more widespread. Today they seem to be mostly confined to West Sumatra and Jambi. Though its range has shrunken vastly and swiftly the creature seems to be hanging on in these areas. We saw the tracks of at least three individual animals in one small corner of Kerinci Seblat National Park, an area that covers 13,791 km^2.

Dr Yanuar examined our cast and concluded that it was an animal track, but it was not from any known animal. He seemed keen that there should be an Indonesian version of my book on orang-pendek. I promised to send him a copy as soon as I got back to England. It would be pleasing to see my work translated and distributed around Indonesia.

During the filming that day there was a slight earth tremor. The ground wobbled for a few seconds but nothing was damaged and no one hurt. Its epicentre was hundreds of miles north in the province of Ache where it killed 35 people and destroyed 4,300 homes.

Whilst online at the hotel, Adele found a news item of extreme interest. Dr Tom Gilbert, a geneticist from the University of Copenhagen, with whom we had worked with before, had begun an exciting new project. Dr Gilbert was looking into extracting DNA from the blood in the guts of leeches to find out what animals they had fed on. Dr Gilbert and his team got twenty five leeches from the Annamite Mountains of Vietnam and successfully extracted mammal DNA from twenty-one of them. Mammals

identified included the striped rabbit (discovered in 1995), the serow, the Chinese ferret-badger and the Turong Son muntjac (that was only discovered in 1998). The DNA can remain intact in the leech's gut for several months. He went on to say that the technique could be used on creatures such as the thylacine and the orang-pendek. We intend to work with Dr Gilbert and his leech method on future expeditions, including our November 2013 expedition in search of the thylacine.

The following day we travelled to Banko. We wanted to find some lowland forest to interview Dr Yanuar in. However the lowland forests were so degraded it proved hard. Most had been chopped down to make way for palm plantations to provide palm oil. We checked into a hotel and then went out to talk to one of the Suku Anak Dalam, the aborigines of Sumatra.

I had met the Suku Anak Dalam twice before on my trips to Sumatra. These people inhabited the island long before the Malays came. Most people in Sumatra these days are Malays. Only 2000 of the Suku Anak Dalam still exist. They are taller and wirier than the Malays with thick, curlier hair. Of course they have a separate language and culture. Until recently they lived wholly in the forest. In more recent days they have taken to living part of the time in houses.

In the past I was told that the Malay name for the 'Kubu' meant 'fortress'. The forest was said to be their fortress. I found out that, in fact, it simply means 'dirty' as most of the Malays look down on the Suku Anak Dalam.

We drove from Banko to a small roadside village to meet a Suku Anak Dalam man called Pak Tumcuggung. Pak Tumcuggung had recently converted to Islam and lived in house eschewing the old ways of his people.

He told us, via a translator, about his encounter. About forty years before, he had been walking through a graveyard about a mile from the village of Batanlumbhi. At the time the area was heavily forested. Today the jungle has been cut down to make way for palm oil plantations. It was the rainy

season and about five in the evening. Pak Tumcuggung came upon some odd looking tracks. Then he saw a man-like, grey coloured figure rise up from behind one of the grave markers. At the time he thought it was a ghost and referred to it as orang-hutan or ghost man.

The creature stood around three feet tall. It had long grey hair, broad shoulders and a pot belly. The face looked very human with broad cheek bones. The creature looked more like an orang-utan than a siamang gibbon. The two stood and stated at each other until Pak Tumcuggung turned and ran. He looked back and saw it still standing there watching him.

At the time he thought what he had seen was a ghost because he saw it in a graveyard and it seemed to combine animal and human features. Now he realises that he saw some kind of animal. He feels it is related to the orang-utan but lives on the ground rather than in trees. He feels it still exists but not in the same area as it has been deforested.

His brother saw an orang-pendek the same year. It was very like the one Pak Tumcuggung described except it had black hair rather than grey. In 2010 local people heard strange calls from an area of forest. It was not a siamang. He thought it may have been an orang-pendek.

The next day we returned to Padang and checked into another hotel. We ate at a branch of Pizza Hut in the city that I always visit after an expedition. We tend to get a craving for western food after a fortnight of Indonesian fare. At least this time around I had avoided rice and spice.

The following day we had an appointment at Andalas University in Padang. Adele had promised to give a demonstration of taxidermy of the students. She showed them how to gut, preserve and stuff a pigeon. I met with Dr Wilson Novarind a biologist at the university. He showed me pictures of tiger, tapir and other animals photographed via camera trap in the jungles that abut the university. The jungle begins literally across the road.

Whilst Adele was giving her demonstration, Wilson and I talked zoology and I mentioned my interest in herpetology. He went on to talk about the records of snake species that had been recorded in and around the university. I asked if any reticulated pythons had been found and he told me of a twelve metre specimen found by people clearing land close to the university. At first I thought he meant twelve feet but he said the snake had been twelve metres or forty feet. The largest reticulated python was thirty-three feet and was shot on the Celebes. This one would have been a full seven feet longer, shattering the record. Wilson showed me the papers that recorded the capture. Sadly there were no photographs. He said that the snake had been sent to a zoo in Bukiittinggi.

I showed him the cast of the orang-pendek track. Like Dr Yanuah, he saw that it was a footprint but could not identify what animal it came from. Wilson said that Cologne Zoo had donated fifty camera traps to the university but they could not afford to ship them over. I suggested that if the CFZ could raise the money to have them sent over some of them could be used to search for orang-pendek in Kerinci Seblet. Wilson seemed very pleased with the idea.

We decided to travel to Bukiittinggi to find the zoo and see if the record breaking snake was still on show there. We had a long drive the next day and found the zoo. I once had a nightmare where I was in a haunted, abandoned zoo where the cages were full of zombie animals. Bukiittinggi Zoo was like that dream come true. It was comparable to the vile Georgetown Zoo in Guyana. Elephants were chained in a tiny enclosure, sun bears walked listlessly around bare enclosures, crocodiles wallowed in shallow pools of oily water, and a single Bornean orang-utan waddled over to paw at us out of the bars. Weirdly the place felt abandoned. There were no keepers in sight. We found an enclosure full of reticulated pythons. None of them exceeded three metres. One would have thought that

the capture of a twelve metre snake would have hit the headlines around the world the way the supposed fifteen metre python supposedly captured in Sumatra in December 2003 did. This monster turned out to be only 7 metres when finally measured in captivity. In 2004 I met one of the men who had captured it. A Suku Anak Dalam chief named Nylam confirmed that it had been seven metres, and the largest he had seen was ten metres. Sadly Nylam himself has now passed away. If there was a giant snake captured in the jungles near Andalas University it seems not to have been on display at Bukiittinggi Zoo. There were no keepers around to ask and the one person we saw there was shutting up a café. He said that he could not recall such a big snake ever being at the zoo. What happened to the giant snake? Did it ever exist? It seems now to be just one of the many mysteries of Sumatra.

Next day we said our goodbyes and headed back to Europe. It had been a pleasure working with Adele and Chris and I can't wait to see the finished film. Chris was very pleased with the evidence we found. I hope to work with them both on future projects.

At the time of writing the hairs are under analysis by Lars Thomas and Dr Tom Gilbert. Adele is repairing the cast. I remain convinced that the orang-pendek is a ground dwelling, bipedal species of orang-utan unknown to science. I think it speciated at the same time as the Sumatran and Bornean orang-utans split off from each other some 400,000 years ago on the landmass of Sunda.

For all its hardships the lost world of Sumatra and the strange creatures and enigmas that haunt it keep calling me back.

The not so mysterious Gardar skull

Lars Thomas

In the summer of 1926 a sensation was dragged from the ground at Gardar in southwestern Greenland. An excavation of an old norse/Viking settlement found among other things a piece of skull and half a lower jaw looking so extraordinary it made international frontpage news, was discussed in the scientific journal *Nature*, and made the leader of the Danish excavation team F. C. C. Hansen describe it as a new species of human that he suggested should be called *Homo gardarensis*. And one has to admit, that especially the lower jaw looks rather strange. The skull is quite thick and solid, and the jawbone is at least twice the depth of a normal one.

Various international experts, though interested in the finds, advised caution. Anthropologist Arthur Keith suggested the bones were from an individual suffering from acromegaly, a glandular disorder resulting in excessive growth of the bones of especially the hands and face. Hansen though was undeterred, and suggested among other things, that if primitive men like this had lived side by side with modern man, they could amongst other things have been the inspiration for the stories about Viking berserker warriors, or perhaps even for the many Scandinavian stories about trolls.

Nevertheless, interest in *Homo gardarensis* started to wane rather quickly, and the specimens eventually ended up in the collections of the Panum Institute in Copenhagen, the University

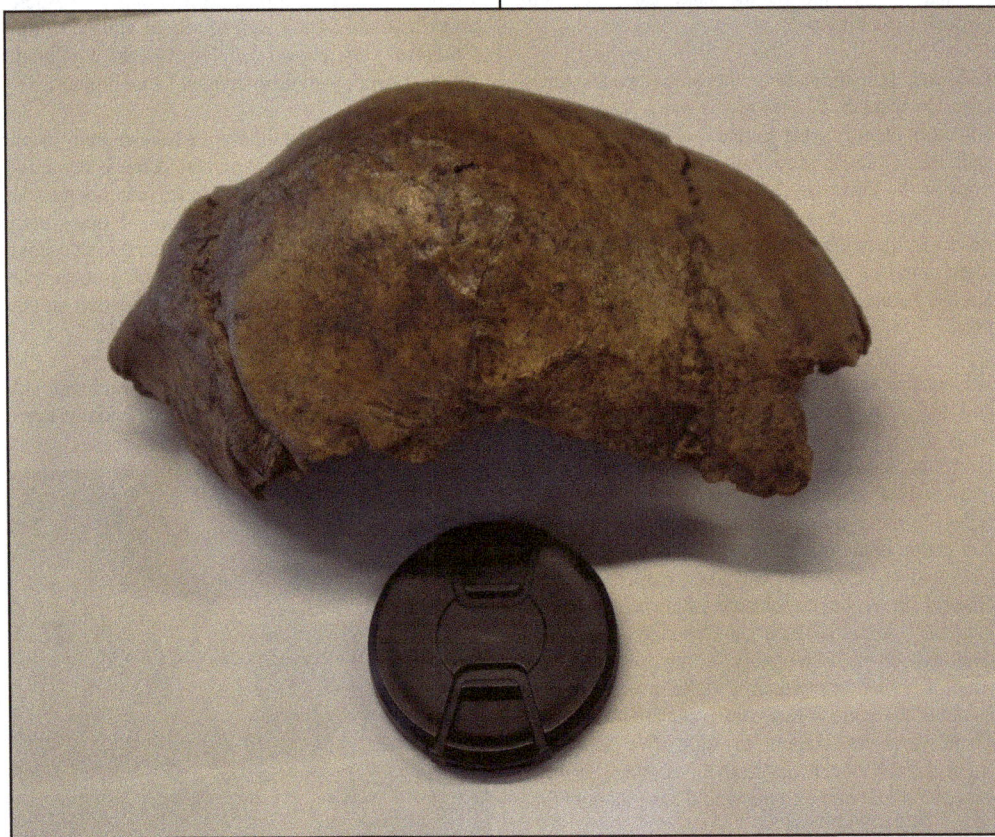

of Copenhagen's department of medicine and anthropology, where they have lain almost forgotten ever since.

ALMOST forgotten that is, because if you take to the internet and start searching for information about the Gardar skull, you will find that a healthy little conspiracy theory seems to have grown up around it. Various writers insist that the skull can't possibly be human, among other reasons because it is HUGE – this idea is to some extent based on Hansen's original attempt to reconstruct the complete skull, as very few of these writers have seen the actual specimens, let alone handled them. And they also insist that the Panum Institute is keeping the skull hidden or simply refuses other scientists and researchers access to them, because if the truth of the skull were to be revealed, all of anthropology would be turned upside down, and all the scientists would sit back with egg on their face, and of course they want to avoid this at all costs. The Gardar skull might even be a specimen of a Bigfoot, alma, almasty – you name it.

Well, as so often before, there is quite a gap between the real world and what has been written about a controversial specimen. When I went to take part in the 2013 Weird Weekend in Devon, to talk about *The Natural History of Trolls* and *The Cryptozoology of Greenland*, I thought it would be nice to bring along some pictures of the Gardar skull – if at all possible.

It turned out to be so easy, it was almost laughable. Professor Niels Lynnerup, who is the curator of the Panum Institute collection of anthropological specimens, was more than forthcoming. Of course I could see the specimens and photograph them to my heart's content. I just had to wear gloves, as sooner or later, as part of the institute's ongoing research in the old Norsemen, the skull will be DNA-tested. So much for the myth of the Gardar skull being hidden and kept out of the reach of researchers.

When Professor Lynnerup then took out a key and opened the glass-fronted cupboard where the skull parts were placed with a copy of the original

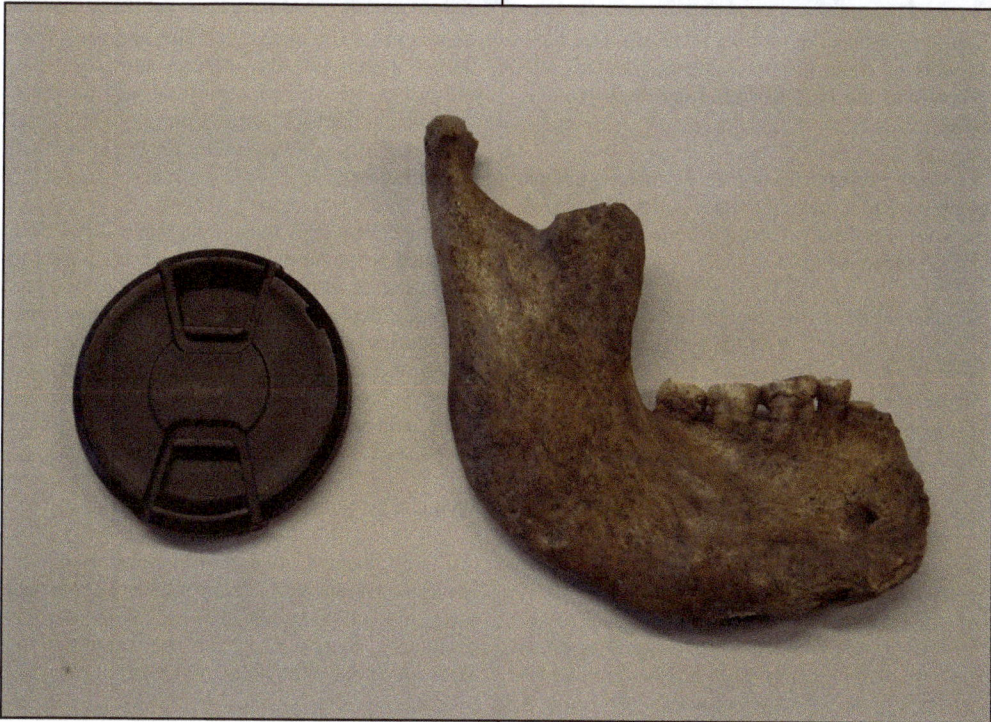

Danish newspaper describing them, my first feeling was of slight disappointment. Was that all? Where was the HUGE skull I had been reading about?

Anyway – Professor Lynnerup showed me into an empty lab, took out a big sheet of paper, placed the skull and the jaw on it, gave me some gloves, and left me to it.

So much for the myth of secrecy.

First of all, the skull is not HUGE, it is well within the size limits of modern man, and as the photograph shows, it actually looks somewhat small. The lens cap in the middle of the photo has a diameter of 65 millimetres, so absolutely no giant here.

This misunderstanding probably arose from the original attempt to reconstruct the whole skull. At the time archaeologists had only rudimentary knowledge of how much bones could be warped, bent, twisted or compressed in the ground, so

what is in fact an almost complete top part of the skull – it ends just above the eyebrows at front, and just above where the junction with the spine would be at the back – was seen as only a partial top part, which gave a skull far too big in the final reconstruction.

The lower jaw does look strange, but it has been flattened in the ground as well, making it easy to reconstruct it as bigger than it actually was. And the large depth of the jawbone is, despite what some writers claim, easily explainable as the result of acromegaly, or even some sort of injury - a severely broken of crushed jaw - and there are some indications of that - which could also have been the cause of the excessive bone growth.

So much for the myth of the giant skull.

All in all, there is nothing mysterious or gigantic about the Gardar skull. It is not being kept locked away, out of the reach of researchers, and it is absolutely not evidence for the existence of Bigfoots, yetis or anything of the sort.

This was very nearly the last *Weird Weekend:* I am getting older, my health is getting worse, and each year there are fewer people that I can rely upon to help. I have been thinking about ending it for several years, and this year I had decided exactly how I was going to do it. I had composed my valedictory speech, and Richard and I had even recorded and filmed our own take on Sid Vicious' version of *My Way* performed by glove puppets.

and now the end is near
and we may face the final curtain
this thing we do each year
the future really is uncertain

There were times everybody knew,
we bit off more, far more than we could chew
for fourteen years, us and our friends,
put on some really weird weekends
will they go on? Well that depends
but we'll do 'em our way

And so on.

In the spring, Dave B-P and I travelled to Brighton where we filmed one of the last performances that Mick Farren did with *The Deviants,* and later interviewed him. It was the first time I had met this remarkable man, to whom I owe so much. If it had not been for his influence upon me I doubt whether I would ever have

This could be The Last Time

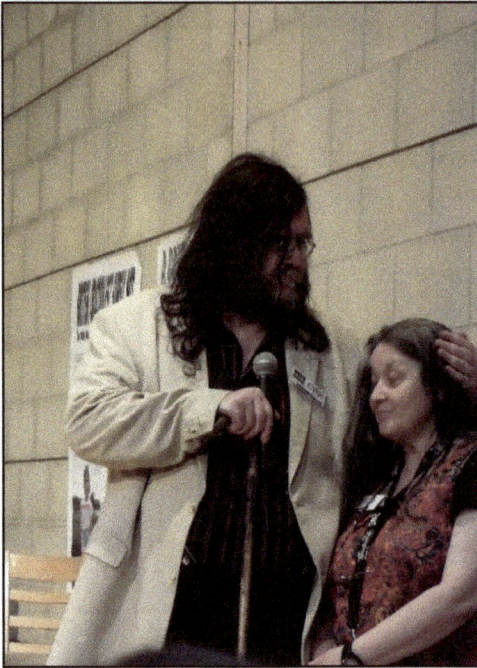

embarked on a career as an independent anarchist publisher, and I am sure that my life would have taken a very different path from the one that it did.

In July, Corinna, my adopted niece Jess Taylor and I were on our way to visit Mick, when the news came through that he had died on stage the night before. He was a great man, and many people (including me) were devastated by the news.

My health is deteriorating, admittedly through many of the same causes that eventually killed Mick, and the stress of the Weird Weekend caused me to have a minor heart episode on the morning of the third day. I had been rehearsing my farewell speech in my head when it happened, and - as clear as day - I heard Mick muttering something unprintable in my inner ear.

Mick died in harness, two songs into a set with *The Deviants* on stage at The Borderline in London, and as I sat gasping for breath with Jess Heard (who took these pictures) holding my hand,

I had an epiphany. Mick lived every inch of his life as if he meant every word that he had ever written. If I was truly one of the people who was taking up his torch, and carrying on his mission to cut through some of the bourgeoisie nonsense that surrounds us all, I could do no less.

So did I deliver my valedictory speech? No, of course I didn't.

In some ways this was the most successful Weird Weekend ever. With a very few exceptions everything went swimmingly, everybody behaved themselves, and a fantastic time was had by all.

This year's speakers were:

- Richard Ingram: *The search for inhabitable planets*
- Ronan Coghlan: *The church and evolution*
- Lee Walker: *Dead of Night*
- Lars Thomas: *The Natural History of Trolls*

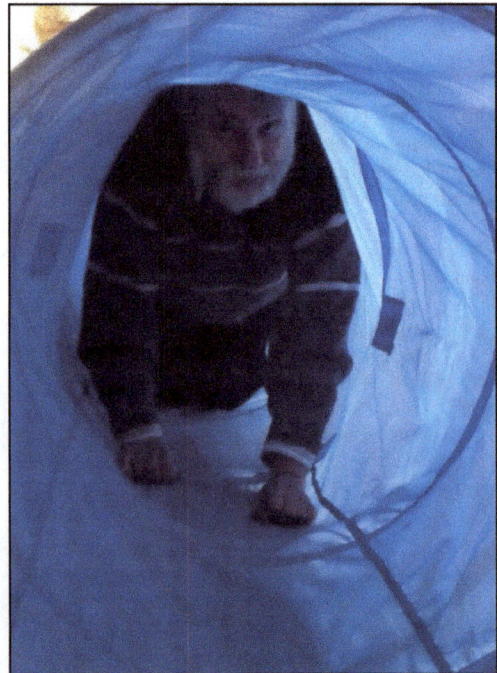

- Jon Downes and Richard Freeman: *Intro to Cryptozoology*
- Nick Wadham: *Fairies*
- Tony Whitehead (RSPB): *Starslime*
- Glen Vaudrey - *Mystery animals of Staffordshire*
- Shaun Histed-Todd: *Evidence for Civilisation X and Pre-Columbian contact*
- Judge Smith: *The Universe Next Door*
- Darren Naish: *Adventures from the world of tetrapod zoology*
- Andrew Sanderson: *Russia 2013 Expedition Report*
- Richard Freeman: *Sumatra 2013 Expedition Report*
- Sarah Boait: *Orbs from a photographer's perspective*
- James Newton: *Bigfoot*
- Lars Thomas: *The cryptozoology of Greenland*
- Ronan Coghlan: *Lure of the Leprechaun*
- Jon Downes: *Keynote speech*

Plus, of course there were the now traditional items of silliness such as Barry Tadcaster and Orang Pendek introducing each of the speeches, the kids' cake eating competition (we, too are worried about the epidemic of childhood obesity, and are determined to do out bit to help), the exceptionally surreal quiz, during which the lovely Kara Wadham wore a pair of fearsome horns (her daughter Lily, another of my adopted nieces, painted her own face blue and wore the horns for much of the rest of the weekend).

Equally surreal was the nature walk which singularly failed to happen because of the rain, but which was replaced by Lars Thomas and Nick Wadham showing a series of pictures of mini-beasts that they had captured in the CFZ grounds during the Thursday evening cocktail party. And of course - for the first time ever - there was Richard Freeman's Tunnel of Goats (in which Ronan Coghlan can be seen on the opposite page).

It is the only cryptozoological gathering in the known universe which starts off with cocktails on my lawn on the Thursday night and ends up with a delicious meal for attendees cooked and served up by the ladies and young people of the village.

I am very pleased to announce that this year, food sales made a profit of £900 for village charities, and that I have already persuaded the lovely Sharon Bennett (who has masterminded this for many years now) to do the same thing next year.

Her words were that if I am not going to give up doing it, then neither will she.

Work is gathering momentum for next year's event. Check out the Facebook Page: https://www.facebook.com/weirdweekend2014 for details.

Tickets will be on sale very soon, and I hereby pledge that the event will continue for as long as I am able to carry on doing it, and - I sincerely hope - for much longer than that.

THE COMMUNITY CENTRE, WOOLFARDISWORTHY, NORTH DEVON, UK   August 15-17 2014
www.cfz.org.uk   TEL: +44 (0) 1237 431413

The editor and his compadres welcome letters for publication on all subjects covered by this magazine. However, we would like to stress that neither this magazine, or the CFZ are responsible for opinions expressed, which are purely those of the letter writer.

First-hand experience of the tokoloshe

EDITOR'S NOTE: In mid-September, I received a telephone call from a lady called Debbie. She was telephoning me because she had read Steven Tucker's excellent book *Terror of the Tokoloshe.* She had a very peculiar story to tell, and so after we had talked for some time, I asked her to write it down...

Dear Jon,

It was a fascinating conversation we had on Friday and as promised, I have written down the personal experience of which we spoke. In some strange way, the story has come full circle and so please bear with me whilst I put the story into context.

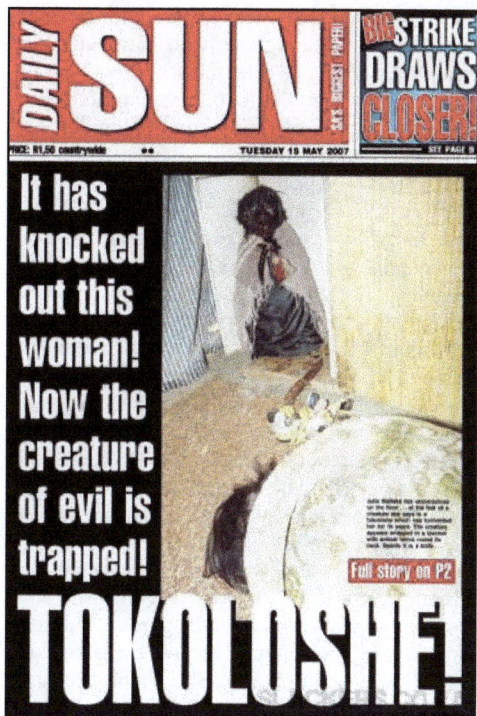

My father, Robert Warne, was born in Pilgrims Rest in Saabe, South Africa in 1914. He was the eldest of five children. My grandfather, Robin Warne, was English and had travelled to South Africa in search of his fortune. It was here that he met my grandmother, Aileen Long, whose family were, to all intents and purposes, South African. My father's own memories of his parents were intensely happy ones, recalling how his parents talked and laughed a great deal together. Tragically, when Dad was about nine years of age, and after his three younger siblings had died (cancer, TB and trying to 'fly' from a top window) his father also died, aged 35. The Warne family decided that Aileen could not cope with her two remaining children and they too were taken away from her. Dad's Uncle George brought him back to England where he grew up at Sherbourne School in Dorset and Midi (his sister) was sent to a convent school in Johannesburg.

Aileen was a witty, intelligent woman. She never remarried and I have recently discovered that for a time, she earned a living by prostitution during which time, she had a further five children. My father did not see his mother for a further 35 years, even though he returned to Africa with his new wife, my mother. They lived and worked in Tanganyika, where I and my two older sisters were born and it was then, on our return from Africa to England for the last time that we met Granny briefly at her home in Saabe.

The events of my life gradually brought me to a point about ten years ago where writing stories about my family and about my ancestors began. The first story was about my father who wanted me to find the whereabouts of a silver rhino that was mounted on a plinth, with the words: *"In gratitude for saving the Earl of Athlone's life."* engraved upon it. It had been presented in gratitude to my Great Uncle George (District Commissioner of Uganda in the 1920s) who had saved the Earl's life after he had taken a shot at a rhino and missed and George had shot it clean between the eyes at it charged towards them and came to rest at their feet. The story was called Chasing Silver Rhinos but I never found the silver rhino and nor did I complete the story.

Since then I have written several stories and then about six weeks ago, having completed and published my last one, I decided that my next project would be to revamp Chasing Silver Rhinos and to turn it into a full, slightly fictitious story. In thinking about the plot, I was keen to incorporate South African folklore, an idea that came from Laurens van de Post's *A Mantis Carol*, the praying mantis being an important god to the South African bushman. It was during the course of researching South African folklore than I came across Mr Tucker's book on the Tokoloshe and when I looked further into this, I was astonished to find the description and uncanny likeness to a real-life event that happened within my own family. It wasn't until the book arrived and that I started to read the Introduction and the fact that there was only one white woman who had experience of the Tokoloshe that I felt compelled to contact Mr Tucker through you. I have yet to read the rest of the book.

So here goes.....

In September 1978, I got married and the following summer of 1979, we went to Skye for our honeymoon. We had booked into the Shligachan Hotel for a few nights and I slept on the right-hand side of the bed, that was closest to the wall, with just a couple of feet between the bed and the wall. I was drifting off to sleep when suddenly I became aware of a small knobbly, black creature/man, about 3 feet high, creeping around the end of the bed towards me. He felt mischievous and slightly sinister and I was intensely aware that he had come to take something from me. I sat bolt upright in bed and shouted, "Oi, no you don't!" This awoke my husband and of course, I had to explain to him what I had seen and felt. My certainty about what had taken place was such that he didn't dismiss the event as a bad dream because the detail was as clear as was the feeling of the creature's purpose. Added to this, my husband

knew very well that I was not a vivid dreamer, nor one of those people who have detailed dreams that they can relay clearly the next day. Unable to form any logic to the situation, it was put to the back of my mind and filed under 'one of those mysteries in life'.

Ten years later, in 1989, my husband and I were living in a place called Mooloolah, near Brisbane in Australia. By this time, we had two daughters, Kathy and Victoria. Victoria was 2 and Kathy was 5 years of age. It had become clear that Kathy had some kind of disability although at that point, no one was sure what had caused her slow development, the obvious thing being that she could only communicate in one-word sentences. It was only later that she was diagnosed with arrested hydrocephalus which had caused partial brain damage at birth. We were renting an old Queenslander and at the time of this second event, my husband was abroad, in Kenya. Kathy and Victoria shared a bedroom which overlooked the garden and their window was open about 4 inches. I was awoken by Kathy at about 5 a.m. She shook me hard and said: "Mummy, a man has been hugging me tightly and breathing very hard on me." I shot out of bed and rushed into their bedroom expecting to find someone but the room was empty. I rushed to the window to see nothing and then realised that, in any case, the window was still only slightly open and certainly not big enough for a man to have climbed out of. In the kerfuffle and confusion, I was aware that my daughter had spoken her first very distinct, well-formed sentence. I was also aware of the fact that Kathy's condition meant that she did not have an active imagination like other children - dolls and teddies were not triggers for imaginative play - and Kathy was incapable of telling lies or embellishment. Calming down a little, I asked Kathy to try and describe the man. She said that he was small, black and bony and that he had hugged her really tightly and breathed heavily on her. I recognised the description instantly.

Not long after this event, my marriage came to an end and I returned to England to live at home with my parents. During this time, Dad

and I would sit up talking about stuff and it was during one of these evenings that I told him of mine and of Kathy's experiences. It was a slight risk on my part because Dad was opposed to flights of fancy but whatever it was that we had been talking about, it seemed appropriate to 'confess' these two strange but unrelated events. Then he looked at me and said that he had never told anyone what he was about to tell me. When he was about five or six - certainly before he had any conscious idea of sex or sexual activity(and long before his father died), he repeatedly had this vision of a little black, knobbly creature/man, getting on top of his mother, and pulling down her knickers to do something that Dad was disturbed by but couldn't put into words.

As in all three instances, and having told the story, I still have no clue to the meaning or to the reason why three generations within one family should have had three similar experiences that were not linked by any previous information or suggestion.

You mentioned the idea of fossil memories i.e. inherent ancient memories that may explain why cultures across the globe have similar mythological creatures e.g. dragons. This is a fascinating idea and in some ways, connects precisely to my own endeavour in life which is to encourage people to write their own ancestral stories - not just for sentimental reasons but because I believe that in the telling of them, we can understand our own inherited fallibilities outside of pure physical attributes. (www.fromthehearth.org.uk). However, in this particular case, the inherited 'memory' seems very specific and very direct.

If you or Mr Tucker can shed some kind of

light on this, I would be very curious to know and I will read Mr Tucker's book with continued enthusiasm!

I look forward to hearing from you in due course.

Debbie Warne

Crypt oturd mystery solved

Hi Jon,

The two-days-per-week job I'm doing at the moment means I can go for a lunchtime walk round the Winfrith nature reserve, where I saw the mysterious "cryptozoological turd" a few years ago (Animals & Men issue 47 page 72). I saw something similar yesterday, in almost the identical location (photo attached). It was similar in size and had the same regular striations, but this one appeared to be fresher (less dried out). I would say that it's fairly clearly a fox dropping, particularly with the distinctive pinched end. What do you think? The scat was surrounded by downy feathers (a few of which are visible in the photo), which again is suggestive of a fox. So if this one is a fox, then the previous one probably was too.

Another question -- is there a way of distinguishing at a glance between a newt and a baby lizard? Both species are supposed to be common in that area, but usually when you see them it's only for a fraction of a second as they scuttle under cover. Is there anything about time of year, location, colouring, way of moving etc that would allow you to tell from a momentary glimpse whether it's a newt or a small lizard?

Andrew May

We seem to be doing something right

Hi Jon,

In the recent "On the Tracks' you've given a lot of time to doing things like removing fish otoliths and different ways of preserving DNA. As much as I enjoy "On the Tracks" in general, I think these have been some of the best segments in a long time. The otolith section especially; I didn't know any of that until Lars explained it.

Sincerely,

Robert Schneck

Cinematic Lexilink

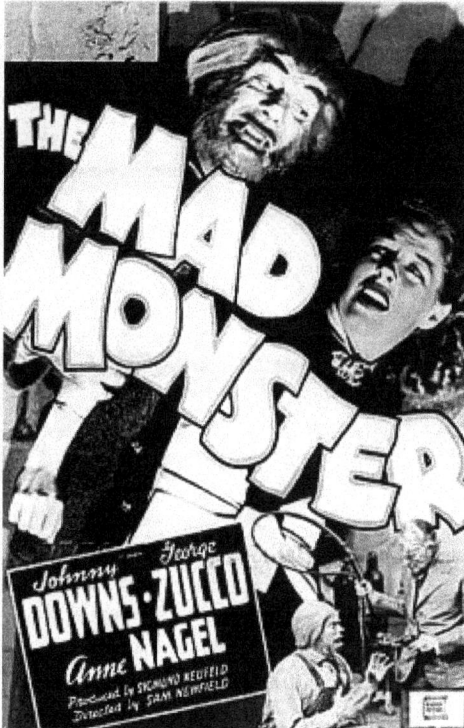

Jonny

U will appreciate the fortean weirdness/"name game" angle of this. I watched an old, obscure monster movie the other night, an old B&W film from decades ago. It was about werewolves. The lead actor was named Jonny Downs!

How weird is that!

Best, Nick

Troll Tails

Dear Mr Downes

The piece by Richard Freeman (page 21 of Animals & Men Issue 50), which referred to similarities in the descriptions of Trolls and the Russian man-beast, the Almasty, struck a chord with me. It reminded me of an article I read in a children's book around 45 years ago. I cannot remember the title of the book but it contained several heavily illustrated chapters on various cryptozoological and ancient historical mystery-type subjects.

I can now only recall one particular chapter, which for some reason stuck in my mind and was entitled something like "The truth about Trolls?" The author claimed that Trolls were in fact a race memory of Neanderthals, which he (or she) suggested may have survived in small groups on the fringes of modern human society for much longer than was thought. Currently, I believe, palaeontologists estimate the extinction date of Neanderthals at around 25,000-30,000 years ago. I think the illustration of the Neanderthal which accompanied the article showed it as a far more savage and ape-like creature than the more recent reconstructions I have seen. Having said that, I imagine it is possible (and I believe others may have suggested this) that the Almasty could be some form of relict Neanderthal or other early human species. After all, the recently discovered Homo Floresiensis (nicknamed "The Hobbit") may have died out only around 12,000 years ago.

Best wishes to all at the CFZ and thank you for a wonderful magazine.

Geoff Clifton

About the Afanc

Hi Jon

Thanks to Bob Skinner, I have been using the Hathi Trust database in recent days to find interesting information of a cryptozoological nature, including stories and accounts of the Afanc. It hasn't cleared up the question as to whether or not the word "afanc" refers to the beaver or crocodile but nevertheless some neglected items have been uncovered.

Rev Roberts *Sketch of the early History of the Cymry, or, Ancient Britons, from the year 700, before Christ, to A.D 500.* page 418 ,thus:

> "The new settlers, under Hu the mighty, are supposed to have retained the traditions of their original country among which the most important is that of the Deluge, which is supposed in the Triads to have happened in his time, or not long previous to it. It is described as having been caused by the bursting of a large lake called Llion, and he is

said by the assistance of his Yehain bannog , or *Buffaloes* , to have drawn an Afanc or *amphibious animal* out of this lake, to have prevented its bursting in future. As this tradition is both curious in itself and important as to a historic fact, it is necessary to consider it with some attention. It cannot fail of striking everyone who is acquainted with the Hindu Mythology; in which, Vishnou destroys the monster who had caused the Deluge, and recovers the Earth and the Veds. It is evidently of the same original, and a part of the general tradition of that aweful event ,which every nation that has ancient records has retained, and applied to its earliest abode after the dispersion, when the memorial was confounded with other emigrations."

From *The Monthly Review; or Literary Journal, Enlarged: Vol XLIV MDCCCIV* (1804) page xxi

> "In Teifi above all the rivers in Wales, were in Giraldus' time a great number of castors, which mat be englished beavers, and are called in Welsh afanc, which name only remaineth in Wales at this day, but what it is

very few can tell. It is a beast not much unlike an otter, but that it is bigger, all hairy saving the tail, which is like a fish tail, as broad as man`s hand. This beast useth as well the water as the land, and hath very sharp teeth, and biteth cruelly, till he perceives the bones crack, his stones be of great efficacy in physic. He that will learn was strong nests they will make , which Giraldus calleth castles, which they build upon the face of the water with great boughs which they cut with teeth,and how some lie upon their backs,holding the wood with their fore feet, etc,etc..."

Source unknown:

"Nant y Afanc , the dale of beavers , is a tremendous vale or rather chasm; the mountains towards the upper part of it approach so near one to the other that they strongly attract the clouds,which ,dissipating around the summits, frequently deluge ,with torrents of rain,the plains below..."

Thomas Roscoe *Wanderings and excursions in South Wales (*1836*)* page 33.

"The beaver was formerly abundant in the Teivy ,though centuries have now passed since its extinction...Old Burton,in his notice of this county (this river forms the boundary between Ceredigion and Carmarthenshire-Richard.) , gives the following account of this curious animal: - "In the river Teivy beavers were formerly found; a creature living both by land and water , having the two fore feet like a dog, wherewith he runs on land ,and the two hinder like a goose with which he swims; his broad tail served for a rudder."

Richard Muirhead

Another fine Ness

EDITOR'S NOTE: In the October 2013 episode of *On The Track* I interviewed David Elder, who - this summer - witnessed something that he could not explain in Loch Ness. You can see the interview on the CFZtv YouTube Channel. I was particularly impressed by the fact that David did not

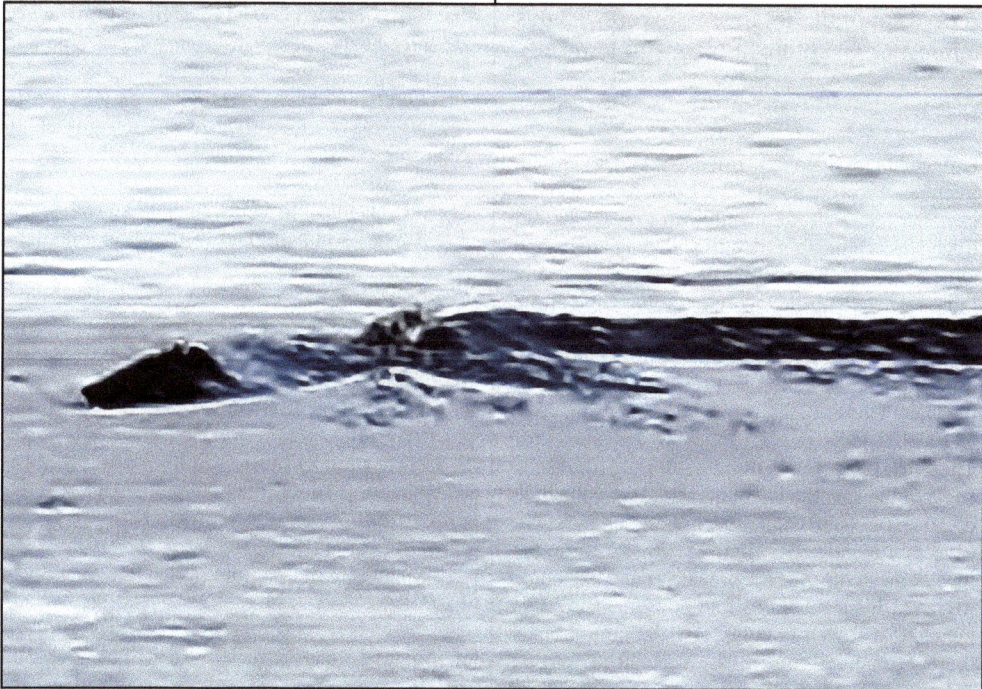

claim that he had seen 'Nessie', but just said that he didn't know *what* he had filmed.

As a result of this interview we received this interesting and thought-provoking letter.

Dear Jon

I watched the video and still shots by David with interest. I go on holiday to Loch Ness, usually camping relatively frequently. In the 90s the old Loch Ness camping park still welcomed tents and myself, my wife and my son went there many times. That campsite sits right on the edge of the loch itself and unfortunately is now a bungalow park. It is about five miles from Fort Augustus, a mile short of Invermoriston on the North side of the loch.

The Loch changes all the time depending on the angle of the sun, the wind, the clouds and the time of day. It is quite common for the Loch to be flat calm as it is in Davids film. If a boat sails up the loch at these times you can watch the wake come to shore, this can take half an hour in some cases but at least fifteen minutes. By the time it lands the boat is long gone. The wake then reflects

back across the loch the other way creating various patterns as the reflected wake hits the wake that hasn't landed. It's very easy to spot a wake if you have sat and watched this. And I fear that this is just what David has filmed. You can see three boats in the background, the two on left look to be heading up the left middle in the direction of Inverness and the one on the right is heading down the right side towards Fort Augustus.

My guess is that attributing animal like features to it is a wee bit hopeful and just wishful thinking.

Where the wake is in the film looks to be around

100 or even 200 yards away from the Fort Augustus pier. The water there is about 300 feet deep. The sides of the loch drop away dramatically and the overall impression is that of a bath tub.

I have an open mind about whether there is a family of large creatures in the loch. I get very frustrated by some of the supposedly 'scientific' explanations though. The current fad I believe is that:

a: there isn't enough food in the loch. Completely wrong, it's stuffed full of arctic char, they are everywhere, it's also got a lot of trout, salmon and eels.

b:It's a sturgeon. Drivel. No doubt the odd sturgeon does find its way to the loch but they are bottom feeders and the loch is vast, unless you are 20 feet from it on a completely calm day you would never spot the back of a sturgeon. You would never spot anything on days when the water is choppy unless it is very near to you or enormous. I once watched a wind surfer from the campsite sail to the opposite shore. His wind surfer sail was not visible to the naked eye when he was half to three quarters across the loch and was quite hard to spot even with 16x binoculars. The loch is about a mile wide at that point.

Another interesting film that has been around for a few years debunks Tim Dinsdales famous film shot from around Foyers. I saw a documentary where they analysed the film and concluded that it was a boat. My recollection, from many years ago now, of reading Nicholas Witchells book on the Loch Ness Monster was that Tim Dinsdale did in fact film a boat for comparison. I'm sure if memory serves me right Witchell includes slides from both the original and the comparison in his book, alas mine disintegrated a while ago so I can't check. In the original film of course the object in the film submerges on the other side of the loch, something which the documentary missed out. I don't know why all this misleading information is out there and I'm a wee bit reluctant to start any new conspiracy theories :)

Any way, tried to leave a comment (three times before I saw the email address)

I visit your website every day, mainly for the big cat stuff as I've had two sightings up here in the Pentland hills. I will get around to joining at some point. Keep up the good work.

Ian Smith

James Herbert Anybody?

Oi Downzy!

Have ya seen this big bugger! Taken from Yahoo:

> Farmers capture one metre-long 'ratzilla' which terrorised village and ate fish WHOLE the freakish giant rodent weighed ten times that of an average rat and had been devouring 3kg fish whole in Shaoyang, in China's Hunan Province.

Davey Curtis

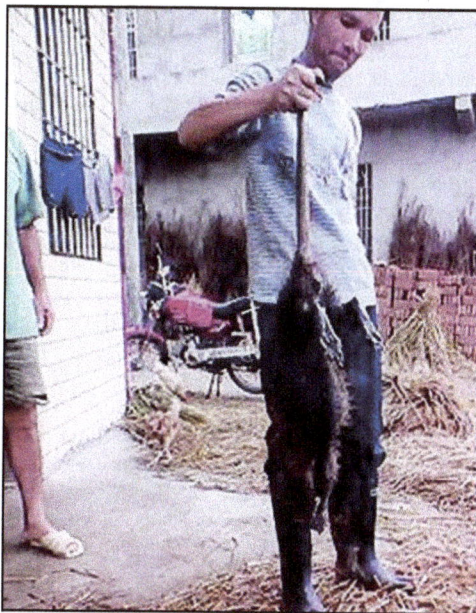

THE WORLD'S WEIRDEST PUBLISHING GROUP

We publish a lot of books. Indeed, I think that we could quite easily claim to be the world's foremost publishers of books about Fortean Zoology and allied disciplines. However, I feel that it would be unethical to review our own titles. So here, to end this edition of *Animals & Men*, is a brief look at some of the books we have put out in the last six months.

- *Mysterious Creatures Vol 1* by George Eberhart

This is probably the title that I am most proud of. We bought the rights to republish it from ABC Clio with the proceeds of the 2013 Weird Weekend, and Volume 2 will be available within a few months. This is a fantastic book, and richly deserves to be the 'Industry Standard' reference book. However, the first edition was marred for many by being prohibitively expensive. Our new edition is channelling the commercial ethics of the anarchopunk which was the soundtrack to much of my youth and is coming out at an easily affordable fifteen quid a volume. I wish it could have been even cheaper.

There are also several hundred new illustrations and the text has been updated wherever necessary. I would like to thank everyone involved in the production, especially George Eberhart for all his input.

- *Mystery Animals of the British Isles: Staffordshire* by Nick Redfern and Glen Vaudrey

The critically acclaimed series of *The Mystery Animals of the British Isles* continues with this fascinating, and witty account of the mystery animals of one of the great historic counties of the Midlands.

Both authors are to be congratulated on the depth of research that they carried out in order to make this a worthy addition to the bookshelf of every Fortean.

- *Terror of the Tokoloshe* by S.D.Tucker

This is another book of which I am very proud, Steven Tucker is an excellent and stylish writer, and this massively researched, and often very funny, account of South Africa's hairy dwarf sex daemon contains an enormous amount of material that quite simply is not available elsewhere.

This is the sort of book that CFZ Press should be putting out; intelligent, academic and wryly subversive tomes that cover areas no other publisher would dare to tread.

www.ingramcontent.com/pod-product-compliance
Lightning Source LLC
Chambersburg PA
CBHW051346290326
41933CB00042B/3303